Building a Sustainable Political Economy: SPERI Research & Policy

The Sheffield Political Economy Research Institute (SPERI) is an innovation in higher education research and outreach. It brings together leading international researchers in the social sciences, policy makers, journalists and opinion formers to reassess and develop proposals in response to the political and economic issues posed by the current combination of financial crisis, shifting economic power and environmental threat. Building a Sustainable Political Economy: SPERI Research & Policy will serve as a key outlet for SPERI's published work. Each title will summarise and disseminate to an academic and postgraduate student audience, as well as directly to policy-makers and journalists, key policy-oriented research findings designed to further the development of a more sustainable future for the national, regional and world economy following the global financial crisis. It takes a holistic and interdisciplinary view of political economy in which the local, national, regional and global interact at all times and in complex ways. The SPERI research agenda, and hence the focus of the series, seeks to explore the core economic and political questions that require us to develop a new sustainable model of political economy at all times and in complex ways.

speri.

More information about this series at
http://www.springer.com/series/14879

Valbona Muzaka

Food, Health and the Knowledge Economy

The State and Intellectual Property in India and Brazil

palgrave
macmillan

Valbona Muzaka
Department of European & International Studies
King's College London
London, United Kingdom

Building a Sustainable Political Economy: SPERI Research & Policy
ISBN 978-1-137-59305-4 ISBN 978-1-137-59306-1 (eBook)
DOI 10.1057/978-1-137-59306-1

Library of Congress Control Number: 2017948160

Cover illustration: Pattern adapted from an Indian cotton print produced in the 19th century

Printed on acid-free paper

This Palgrave Macmillan imprint is published by Springer Nature
The registered company is Macmillan Publishers Ltd.
The registered company address is: The Campus, 4 Crinan Street, London, N1 9XW, United Kingdom

Acknowledgements

My interest in India and Brazil was sparked during my earlier research on international negotiations and contestations over intellectual property. I wanted to understand what lay behind their international activism in this aspect of global governance, a curiosity that has only been very partially satisfied during the writing of this book. I say partially advisedly, aware of the complex realities of India and Brazil, to whose understanding many of my peers have dedicated their professional lives. Besides my curiosity, giving back to our common pool of knowledge something in return for the great insights I have drawn from it has been another motivation while writing this book. That this is now a book is something for which I cannot claim all the credit. I say this not only because any book—as knowledge itself—is always the outcome of collective efforts. I say it especially because it was unlikely to have been realised in its current form without the gentle push by this series' editor, Tony Payne. He and Colin Hay have been supportive throughout and I would like to thank them for believing in this project as much as did. I would also like to thank the Leverhulme Trust for finding the original project interesting and worthy enough to support it financially (RF-2012-060). Its support allowed me to spend time carrying out research and interviewing various stakeholders in India and Brazil. My thanks go unreservedly to the many people who freely gave their time to discuss with me various aspects of their work related to the ongoing intellectual property reforms in both countries. A fellowship grant from CONFAP–British Academies in 2016 was helpful in enabling me to spend more time in Brazil. I would particularly like to thank colleagues at the Department of Economics at the Universidade Federal de Minas Gerais in Belo Horizonte, with whom I had the pleasure to spend four months at the end of 2016. Their generosity has been unsurpassable, but no one was more generous with his time, intellectual advice and support of all kinds than Eduardo da Motta e Albuquerque. From him I received much, perhaps above all the pleasure of enjoying the extraordinary combination of intelligence, integrity, humour and generosity that he possesses. I would also like to thank my reviewer, who read the book as critically and constructively as anyone could ever wish for from a reviewer. It not only made this book better, but, if I manage to apply what I learnt from him, I fancy that it may make me a better reviewer, too. For their many efforts and enduring support throughout the process, I should also like to thank Christina Brian and James Safford at Palgrave. I am also indebted to my colleagues at the Department of International and European Studies at King's College London for their flexibility and support shown during the time it took me to bring this book to fruition. Special thanks go to my siblings—Anila, Plaurent, Rovena and Alban—and their families, who are for me a never-ending source of love, humour and joy. The most heartfelt gratitude goes to Richard, not only for reading absolutely everything I wrote multiple times and for being saintly patient towards the stubbornness with which I often met his helpful advice, but above all, for the wonderful gift of himself that he freely gives and I daily enjoy.

Contents

CHAPTER 1

Introduction

Ongoing methodological disputes notwithstanding, there is little doubt that intellectual property (IP) markets have grown substantially since the 1980s. By some accounts, total receipts for the use of IP worldwide were estimated to have reached US$ 318 billion in 2015, up from US$10 billion in 1980 and just under US$3 million in 1962 (current US$).[1] Despite the increasing internationalisation of research and development (R&D) activities, around 88% of income from IP titles such as patents, copyrights and trademarks (though not tax extracted on such income[2]) continues to flow to the US, the EU and Japan. Of the many factors contributing to this heavily skewed distribution of IP rents, the concentration of high-tech sectors in the hands of a few market actors headquartered in the wealthy regions of the world economy is crucial. For instance, around 44% of the global pharmaceutical market, 50% of the global seed market and over 80% of the pesticide one are controlled by the top ten companies in each respective sector (Kesič 2011: 218; Brand et al. 2008: 19), almost all of which are headquartered in frontier economies and use IP titles intensively as a market control and monetarisation tool. Their advantageous market position is based to a significant extent on the structural advantages of frontier economies, the most notable of which is perhaps the relatively strong R&D base developed over time with substantial contributions from public sources: from the end of World War II (WWII) until the 1980s, the state funded more than 50% of all R&D in the OECD (Organisation for

V. Muzaka, *Food, Health and the Knowledge Economy*, Building a Sustainable Political Economy: SPERI Research & Policy, DOI 10.1057/978-1-137-59306-1_1

Economic Co-operation and Development) members (Niosi and Faucher 1991: 133). A whole raft of other measures undertaken and underwritten by states have further enhanced the capabilities of these business actors to control markets and enhance incomes appropriated from them. One important way in which frontier states have attempted to strengthen the position of their high-tech sectors in global knowledge markets has been that of expanding the remit and the temporal and spatial reach of IP titles. The transformation of collective social knowledge into private knowledge via the institutional form of IP—a precondition for the existence of knowledge and IP markets and income appropriated through them—would not have been possible were it not for the state acting as the guarantor, enforcer and legitimator of IP titles. In other words, the state plays a central role in creating and sustaining IP markets. Moreover, the state is not only a market maker but also a market player. State agencies in frontier economies, for instance, are important players in IP markets themselves, for example, as owners of IP titles or, to offer a recent example, of the so-called sovereign patent funds that act as state-backed investment vehicles to encourage the expansion and deepening of patent markets.

The notable expansion of IP markets alongside the even more remarkable expansion of financial markets is a manifestation of deeper tendencies that have come to characterise the more advanced regions of the global economy from the late 1970s onwards. Perhaps the more notable of these has been a shift in the locus of wealth creation away from productive activities towards wealth created through the manipulation and exploitation of knowledge and information, as well as through rents on assets such as financial and IP titles. 'Financialisation' and 'the knowledge economy' are two frequently used, if often vague, terms that capture these tendencies. Usually studied separately, these two tendencies are in fact closely intertwined. But contrary to the generally more critical treatment of financialisation and its consequences, the knowledge economy has many vocal supporters, of which the most notable are states themselves. Often invoking the difficult hand dealt by financial markets to justify an unpopular move, state representatives are just as often found praising the advantages of the knowledge economy. Since this particular way of conceptualising and organising socio-economic activities became dominant in frontier economies in the early 1990s, states everywhere displayed no immunity to its charms: it is almost impossible to find official policy announcements that do not make reference in one way or another to the necessity and advantages of becoming a knowledge economy or participating in the global knowledge economy today. As it (and more recent mutations such as bioeconomy, gig economy, digital economy or network economy) became the preferred way of conceptualising the economy and its future direction in the more economically advanced countries, our search for understanding its nature and exigencies also followed the same trajectory, generating a body of work that is largely focused on frontier countries, while neglecting others who, by choice or necessity, are attempting to make the leap into the global knowledge economy.

The most notable 'others' in this context are China, India and Brazil. As domestic and international pressures led the state in all three countries to contemplate a new growth strategy by the late 1970s and early 1980s, the decisions taken, although distinctive, were unmistakably orientated towards transforming these economies into competitive knowledge-based economies. In China, for instance, Deng Xiaoping's Four Modernisations strategy—modernisation of agriculture, industry, science and technology, and defence—was predicated to a significant extent on science providing the route to 'socialism with Chinese characteristics' (Keeley 2005). The aim of the ensuing 863 National High-Tech Programme established in 1986 was to enable China to co-opt the new sciences (e.g. information technology, new materials and biotechnology) before they came to be dominated by the West and the US in particular (Jakobson 2007). Enabling China to operate close to the technological frontier was seen as a matter of not merely economic survival in the changed circumstances, but also of national pride: China finally *re*gaining its rightful place in the world following centuries of humiliation and semi-colonial rule (Hsü 2000). As increasing funds were channelled towards financing R&D—reaching the EU level of around 2% of gross domestic product (GDP) in 2015[3]—the state's vision of transforming China into a knowledge economy included the ambitious goal of reaching the status of 'innovation-orientated' country by 2020 and a world leading science power by 2050 (Jakobson 2007). Following a different route, the Indian state, equally intent on re-establishing India as a world power, made clear that it was working towards transforming India into an innovative economy and one of the top five global knowledge powers by 2020 (GoI 2013). Like China and India, Brazil has made no secret of its ambitions to move from 'emerging power' to 'knowledge power' status. Compelled on the one hand by changes in the global economy and on the other by its own version of exceptionalism, it, too, has been attempting the transformation of its socio-economic structures with a view to becoming the great technological and environmental power of the twenty-first century (Lula 2007).

Evaluations based on indicators such as gross R&D expenditure as a percentage of GDP or the average rate of annual growth in R&D spending would decidedly put China ahead of India and Brazil. But focusing on India and Brazil, as is the intent here, is arguably more rewarding for a number of reasons. First, since the late 1940s, the state in both India and Brazil has shown a consistent commitment towards supporting R&D and investing in the nation's science and technology base as a key element of its chosen growth strategy, a consistency that the Chinese state cannot claim. Second, China turned its attention seriously to IP issues domestically during the 1990s and only found its voice in international IP fora when the first decade of the twenty-first century was coming to a close (Yu 2013). By contrast, India and Brazil both inherited IP statutes dating back to the 1800s and, more importantly, this experience with domestic IP rules was accompanied since early on with activism in international fora. As will be seen in Chap. 4, the Indian and the Brazilian states have a long record of involvement with IP issues internationally, more than any other ambitious contenders, starting from the 1950s and continuing to this day. The fourth chapter provides an account of these IP contests, focusing in particular on the most significant conflicts related to (pharmaceutical) patents and plant genetic resources up until the early 1990s, when the consolidation of the knowledge economy imaginary in advanced economies was accompanied—not by chance—by the arrival of a global and much stricter IP regime that was ushered in by the 1995 WTO TRIPS Agreement.[4] It is on account of this activism that India and Brazil have earned the label of 'leaders of the developing world' on IP issues, a label China is yet to attain. Third, China's success is often presented—although not very convincingly—as an example of a state that understands the exigencies of the knowledge economy and responds to them effectively and robustly, untroubled and unhindered by social conflicts. This picture, simplistic as it is, contrasts strongly with the 'slower' progress made by the Indian and Brazilian states on account of myriad economic and political impediments and all manner of conflict. These impediments and conflicts are precisely why understanding the role the state in both has played in shaping IP rules, and the tensions that have unfolded as a result, may prove a more interesting and rewarding exercise.

It will be said that these reasons appear justifiable only insofar as a focus on the state and its role in shaping the IP institutional form is justified. Why explore the many exigencies the knowledge economy places upon a society through the prism of the institutional form of IP? As will be discussed at more length in the second chapter, what sets the knowledge economy apart from other socio-economic formations is not that the latter were not knowledge based, but that now knowledge functions more like a fictitious commodity and fictitious asset than in these other formations. The primary institutional form that mediates the functioning and circulation of knowledge as a commodity and asset is that of IP. Commonly referred to as intellectual property rights—a term avoided here as the language of rights naturalises the many problematic aspects of this institution—IP is an umbrella term that includes within its remit a disparate and ever-expanding set of titles of which the most familiar remain patents, trademarks and copyrights. From its rudimentary beginnings in the fifteenth-century Europe, the IP institution has taken different forms across time and space, but it could be said that, especially from the 1980s onwards, it has more or less eliminated alternative ways of organising knowledge production and circulation in favour of a scarcity- and market-based model that has increasingly come to cater to private-market actors' short-term interests. None of this could have occurred without (the frontier) states creating, guaranteeing, enforcing and expanding IP titles in ever larger parts of the world and for longer time periods on behalf of these actors.

If this account strengthens the rationale for focusing on IP in the context of the knowledge economy, objections may still be raised about the proposed focus on the state. Why this focus given the many empirical and theoretical observations made in various disciplines about the necessity of identifying conceptual units that are better able to capture the fluid and complex nature that characterises the current processes of wealth creation, circulation and appropriation? The second chapter also provides a more detailed account of why focusing on the state is still relevant today, despite the irrefutable evidence of radical changes in the global structures of production, trade and finance, and the attendant dethroning of the national/domestic as the most relevant unit of analysis. Focusing on some of the key trends that have informed such changes in the context of the shift from Fordism to post-Fordism in advanced economies, the argument is made that, although the national space appears to have lost its primacy, no other new scale of economic and political organisation has usurped it; in other words, the unique socio-spatial location of the modern state has not yet been occupied by another actor (Hobson 1997; Jessop 2000). As will become clearer, far from withering away, the multifaceted exigencies of new forms of competition that have come to characterise post-Fordism require the state to attend to an expanding list of tasks: increasing labour market flexibility, reforming the terms of pre-competitive firm relations, increasing investment in science, reforming the role of universities, creating new markets, changing the terms of finance's engagement in knowledge markets, regulating genetic resource use, reforming IP rules and, importantly, cultivating consensus for the new growth regime and the many changes deemed necessary to attain and sustain it.

The institutional forms thought to be necessary to succeed in the global economy do not appear automatically, but are actively formed and reformed by social actors, among which the state plays a commanding role, even if its own form is changed in the process. The terms under which these processes are occurring under post-Fordism are specific to it, but state forms have never been static; rather, reflecting states' dual anchorage in the domestic and international level, state forms have always been historically specific expressions of conflicts among domestic groups and, simultaneously, of the workings of the world economy (Hobson 1997; Cumings 1999). This is why institutional forms change between states and across time, and why the mode of analysis adopted here seeks to retain a dual focus on both major changes in the world economy and the manner in which these have played out in conflicts between social groups domestically in India and Brazil. One of the main reasons for discussing the most important changes to the terms of competition in the world economy in the second chapter is to delineate the way in which the institutional transformation they set in train in the more economically advanced countries were not limited to them but provoked a series of reforms in other countries. Pressures were perceived and articulated differently in different domestic contexts, but, as the analysis in Chaps. 5 and 6 on India and Brazil (respectively) indicates, the state in both—or, more accurately, important fragments within it—felt compelled to embrace the new knowledge economy and competitiveness orientation as the state in more advanced countries had done. This change in orientation, in turn, required considerable institutional changes, the most important of which for our purposes are those to the institution of IP. As the analysis in the following chapters makes clear, the performative and constitutive role of economic imaginaries such as that of the knowledge economy does not unfold automatically, but is rather fomented in social conflicts that become the more pronounced the more radical the changes demanded by the new orientation.

Interrogating the role of the Indian and the Brazilian state in the context of the knowledge economy is not meant to legitimise and naturalise the knowledge economy and all its attendant exigencies. Efforts are made in the second and third chapters to reveal the origins of the knowledge economy as an economic imaginary politically supported by particular groups and whose materialisation and consolidation in practice is by no means taken to have been completed in frontier economies or elsewhere. Interrogating the measures taken by the Indian and the Brazilian states is neither an endorsement of such measures and of their apparent necessity for propelling these economies towards a knowledge economy status, nor of the desirability of such status. Rather, the focus here is on analysing why the state in both countries embraced the imaginary of transforming their economies into knowledge economies able to compete in the global market, and what role it played in the multiple social conflicts that this orientation occasioned particularly in the area of IP. While IP and its effects on the distribution of social wealth have everywhere made it a central sphere of social and political conflict, the focus here is on IP titles most relevant to two specific sectors: pharmaceutical and agro-biotech. These two sectors are chosen for a number of reasons, the most important being that they are considered by both India and Brazil as high-tech sectors whose development is key to mastering competencies needed to operate at or near the technological frontier and, more broadly, to nudging their economies towards a technologically advanced, competitive knowledge economy status. Moreover, both sectors come under 'life science industries', which since the 1990s have taken over the chemical and atomic research/sectors as principal national priorities in most frontier economies and arguably worldwide. In addition, these two sectors are important not only because the Indian and the Brazilian states have come to perceive them as central to economic competitiveness and growth, but also because of their contribution to meeting human needs—primarily food and health—that are still rather pressing for a considerable part of the population in both countries. For these reasons and more besides, developments in these sectors are a focal point where social and political conflicts have converged in India and Brazil and in which the state, as will be seen in the following chapters, plays a key role.

Ever since the discovery of the structure of the DNA in a university lab in 1953 opened the way to an impressive range of developments in genetics and molecular biology, red and green biotechnologies (biomedicines and agri-biotech, respectively) have generated considerable hopes for business actors and states alike. For market actors, biotechnology opened a new frontier that transformed life in its multiple forms into a central locus of economic value creation and extraction. For their part, states became major supporters of biotechnology because it offered not only hopes for economic regeneration and competitive advantage in world markets but, as many problems could be reformulated so as to take advantage of biotechnology, also hopes of strengthening their own legitimacy (Jasanoff 2005). Although expectations of new technologies often reach hyperbolic levels in the early stages of their emergence, it could be confidently argued that hopes pegged on life sciences remain undimmed today, despite the fact that products have not yet reached markets in large numbers and the vast majority of biotech companies' profits—despite expanding revenues—hover consistently at or below zero (Pisano 2006). As will be discussed in the third chapter, this state of affairs has much to do with the manner in which biotechnology as a commercial sector developed in the US, initially, and later in other countries. This discussion is also important because in seeking to become competitive knowledge economies—a hope that for India and Brazil hinges to a significant extent on life sciences in general and biotechnology in particular—both states have found inspiration for what they deem necessary institutional reforms in those undertaken initially by the US state, even today the undisputable leader in this field.

Many factors have contributed to this leadership position, but, importantly for our discussion, it could not have been achieved or retained in the absence of the new global IP regime ushered in by the WTO TRIPS Agreement. As will be seen, being primarily a means of protecting the competitive position of the high-tech (especially life sciences) sectors of advanced economies and expanding their technological rents, TRIPS poses significant challenges to developing countries that, like India and Brazil, hope to become key players in the knowledge economy. At least as far as IP reforms are concerned, it could be argued that India and Brazil could not have been innovative given the restrictive and legally binding changes to the global IP regime ushered in by TRIPS. Besides, they resisted its negotiation at the start and have continued to contest various aspects of the global IP regime after TRIPS entered in force. There is truth in these assertions, but their final acceptance of TRIPS and subsequent IP reforms would likely not have occurred or succeeded had the state in both not adopted the competition state orientation, discussed in more detail in Chaps. 5 and 6. The most important (international) political conflicts over IP spearheaded by Brazil and India during the so-called IP 'Dark Ages', and the ways these were superseded by TRIPS are discussed in Chap. 4.

Revisiting this turbulent prelude to TRIPS in this chapter helps embed the adoption of the new competitiveness orientation on the part of India and Brazil and concomitant changes to their domestic IP systems in a broader historical perspective. It also serves the purpose of sketching the trajectory of contests over IP for pharmaceuticals and plant genetic resources that, having hitherto unfolded largely separately, following developments in biotech came head to head in the late 1980s with the negotiation of TRIPS and of the Convention of Biodiversity (CBD). Far from resolving conflicts, these two key developments simply opened up new international fora where states preoccupied with their competitive position continue to this day to attempt shaping the rules of the game in favour of sectors they deem most important to their growth and competitiveness.

As the analysis of the measures taken by the Indian and the Brazilian states regarding pharmaceuticals and agro-biotech suggests, both have been among the most active players challenging the increasing commodification of knowledge and of genetic resources in international fora on the one hand, while pursuing such commodification domestically on the other. Not unlike their counterparts in frontier economies, state agencies in Brazil and India are also key players in domestic knowledge markets; although they have not yet created vehicles such as sovereign patent funds, state agencies are by far the most aggressive domestic knowledge market players in both India and Brazil. Embedding the analysis throughout in a conceptual framework that considers the state to be simultaneously a domestic expression and shaped by the exigencies of international competition, the ambitions and the activities of the Indian and the Brazilian states are investigated in light of the historically distinct manner in which the knowledge economy imaginaries developed domestically from the 1980s onwards, and in light of the continuities and departures these imaginaries occasioned on the role the state had played in negotiating the space between the national and the global in the past. Based on this mode of enquiry, the analysis offered throughout and brought together in the concluding chapter is concerned less with offering a comparative account between India and Brazil and more with investigating the manner in which similar pressures gave rise to local conflicts whose outcomes inevitably bear the marks of the distinctive and differently constituted terrains in which they unfolded. Looking ahead, the concluding chapter also makes an attempt to tease out some of the main challenges India and Brazil face in realising their hope of becoming competitive knowledge economies in the twenty-first century.

Notes

1. World Bank, World Development Indicators database, available at http://data.worldbank.org/indicator/BX.GSR.ROYL.CD, last accessed 5 April 2017.
2. For an account of the multiple ways in which tax on intangible assets like IP rents is avoided, see Bryan et al. (2017).
3. See OECD data, available at https://data.oecd.org/rd/gross-domestic-spending-on-r-d.htm, last accessed 5 April 2017.
4. TRIPS stands for Trade-Related Aspects of Intellectual Property, a euphemism to justify the inclusion of IP under the World Trade Organization (WTO) remit, since all IP known and in use at the time were included in the agreement.

Bibliography

Brand, U., C. Görg, J. Hirsch, and M. Wissen. 2008. *Conflicts in Environmental Regulation and the Internationalisation of the State: Contested Terrains.* London: Routledge.

Bryan, D., M. Rafferty, and D. Wigan. 2017. Capital Unchained: Finance, Intangible Assets and the Double Life of Capital in the Offshore World. *Review of International Political Economy* 24 (1): 56–86.

Cumings, B. 1999. Webs with No Spiders, Spiders with No Webs: The Genealogy of the Developmental State. In *The Developmental State*, ed. M. Woo-Cumings. Ithaca, NY: Cornell University Press.

GoI. 2013. *Science.* New Delhi: Technology and Innovation Policy, Ministry of Science and Technology.

Hobson, J. 1997. *The Wealth of States: A Comparative Sociology of International Economic and Political Change.* Cambridge: Cambridge University Press.

Hsü, I.C.Y. 2000. *The Rise of Modern China.* Oxford: Oxford University Press.

Jakobson, L. 2007. China Aims High in Science and Technology: An Overview of Challenges Ahead. In *Innovation with Chinese Characteristics: High Tech Research in China*, ed. L. Jakobson, 1–36. Basingstoke: Palgrave Macmillan.

Jasanoff, S. 2005. *Designs on Nature: Science and Democracy in Europe and the United States.* Princeton, NJ: Princeton University Press.

Jessop, B. 2000. The State and Contradictions of the Knowledge-Driven Economy. In *Knowledge, Space, Economy*, ed. J.R. Bryson, P.W. Daniels, N.D. Henry, and J. Pollard, 63–78. London: Routledge.

Keeley, J. 2005. Interrogating China's Biotechnology Revolution: Contesting Dominant Science Policy Cultures. In *Science and Citizens*, ed. I. Scoones, M. Leach, and B. Wynne, 155–166. London: Zed Books.

Kesič, D. 2011. Strategic Development Trends in the World Pharmaceutical Industry. *Managing Global Transitions* 9 (3): 207–223.

Lula (Luiz Inácio da Silva). 2007. Discurso do Presidente da República, Luiz Inácio Lula da Silva, na cerimônia de lançamento da Política de Biotecnologia, Palácio do Planalto, Brasília, 08 de fevereiro de 2007. Available at http://www.biblioteca.presidencia.gov.br/presidencia/ex-presidentes/luiz-inacio-lula-da-silva/discursos/

Niosi, J., and P. Faucher. 1991. The State and International Trade: Technology and Competitiveness. In *Technology and National Competitiveness: Oligopoly, Technological Innovation, and International Competition*, ed. J. Niosi, 119–141. Montreal: McGill-Queen's University Press.

Pisano, G.P. 2006. *Science Business: The Promise, the Reality and the Future of Biotech*. Boston, MA: Harvard Business School Press.

Yu, P. 2013. The First Decade of TRIPS in China. In *China and Global Trade Governance: China's First Decade in the World Trade Organization*, ed. K. Zeng and W. Liang, 126–143. Abingdon: Routledge.

CHAPTER 2

Catching Up in the New Knowledge Economy

Of the successful 'catch-up' stories of the nineteenth century, it was in Germany that *staatswissenschaft*—the science of the state—developed, a political theory of late development that would be embraced to varying degrees by all late industrialisers. The central role granted to the state in achieving socio-economic development in theory and practice from then onwards was helped in no small measure by the way the nascent discipline of economics came to define the national economy with fixed geospatial boundaries as a measurable, manageable and manipulable economic unit (Bergeron 2004). But whatever latecomers could learn by observing how certain states had successfully managed the development of their national economies provided only limited guidance, for, as Gerschenkron (1962) pointed out, their field of action was a changing and challenging one. Every successful 'catch-up' case transformed to some degree the nature of the world economy, raising the bar for other states attempting the leap and necessitating ever newer and more complex strategies and institutional innovations in order to succeed. Indeed, seen from the distance afforded by historical perspective, in the most successful late developers of the nineteenth century—Germany, the US and Japan—the state had played a relatively 'simple' role when compared with the developmental states of the twentieth century, whose activities extended to direct and major investments in infrastructure and industry, the regulation and monitoring of private investment and activity, investment in science and education

V. Muzaka, *Food, Health and the Knowledge Economy*, Building a Sustainable Political Economy: SPERI Research & Policy,
DOI 10.1057/978-1-137-59306-1_2

and so on (Chibber 2005). In turn, the achievements of the most successful developmental states of the twentieth century—Japan, South Korea, Taiwan and Singapore—once again complicated the task of ambitious states that had remained behind as the new century approached.

The challenges that face the developmental states of the twenty-first century stem not only from the necessity of designing new policies and institutional forms in the wake of successful catch-up strategies adopted elsewhere that, to make matters worse, offer them only limited guidance for the task ahead. Were attention to be paid only to the transformations brought about by such strategies—a concrete instance of which is the current debate on the difficulties the rise of China poses to other developing states seeking to strengthen manufacturing—a rather truncated understanding of developmental challenges would be achieved. While certain previously marginal spaces have more recently acquired enough power to influence the general dynamics of capitalism,[1] it still remains the case that the most radical transformations continue to emanate from the core of the global economy. This is to say that the challenges faced by states seeking to catch up in the twenty-first century stem not only from transformations brought about by earlier catch-up successes, but also—indeed, primarily—from broader transformations whose origins are to be found in frontier economies. As capitalism spread out, from these economies emerged pressures that shaped and redefined socio-economic realities elsewhere in ways contingent on prevailing socio-economic–historic contexts, and on these economies the gaze of catch-up contenders remains fixed.

The most important transformation to have occurred in frontier economies in recent times is what the regulation approach refers to as the shift from Fordism to post-Fordism. The brief discussion of this shift in the following section is not offered as evidence that post-Fordism is now a clearly defined, stable or uniform formation at the core, but rather to point out that the transformations it has set in train in frontier economies are by no means limited to them. The state—anywhere—has historically been formed and transformed by international competition in the world market (Cumings 1999: 92), but the kind of competition instituted by the shift to post-Fordism in the context of a more open global economy has unleashed a rather more forceful 'whip of external necessity' (Selwyn 2011) on countries seeking to catch up in the twenty-first century. This warrants, in turn, an interrogation of the key characteristics of post-Fordism. For our purposes, the most relevant insights can be gleaned from work on the shift from the Keynesian Welfare State to the

Schumpeterian Workfare State orientation (Jessop 1993, 2000) and the rise of the internationalised competition state orientation (Cerny 1997; Brand et al. 2008), alongside concomitant socio-economic transformations implicated in this shift. Of the latter, the second section focuses on the emergence of the so-called knowledge economy from the late 1970s onwards. The knowledge economy imaginary constitutes one of the most powerful discursive and material forces that have come to orient not only the policies and strategies of frontier economies, but, irrespective of its merits for the task ahead, also of ambitious catch-up countries such as India and Brazil. While the features of the knowledge economy are yet to become fully clear, it is not hasty to argue that of the multiple and complex challenges that it poses to aspiring contenders, the most important relate to how the 'new capital'—knowledge—is owned, accumulated and circulated. The enabling and conditioning elements of knowledge functioning as a commodity and as capital are largely determined by the domestic and global arrangements that govern intellectual property (IP), discussed in the final section of this chapter.

2.1 From Fordism to Post-Fordism: The Emergence of the Internationalised Competition State Orientation

Since its emergence during the 1970s and drawing from a number of scholarly currents that include Marxist, post-Keynesian and institutionalist theories, the regulation approach has sought to provide a framework aimed at understanding, on the one hand, the historically and spatially differentiated manner in which social institutional forms can provide a relative stability for a specific regime of growth, and, on the other, the displacement of such a relatively stable formation by another one in the wake of secular crises (Jessop 1997; Brand et al. 2008; Becker et al. 2010). The main premise of this approach is that capitalism is not a self-equilibrating process, and that a certain configuration of sociopolitical institutional forms mediates, regularises and stabilises a particular growth regime in a distinct historical phase and place (Jessop 1997). The main institutional forms that dominate regulationist analyses are the state, money form, inter-state relations, forms of competition and wage relations (Boyer 1990; Petit 1999). These institutional forms are not strictly juridical or economic; they include both economic and non-/extra-economic mechanisms that are coupled to and stabilise a particular growth model. Historically, the state has played a central role in shaping the other institutional forms, a role that, as will be discussed, has not disappeared. From this perspective, Atlantic Fordism was a growth model that was mediated by an institutional context patterned by the (differently expressed) Keynesian state. While actual arrangements varied in different countries, state intervention tended to be most active in the institutional form of wage labour and employment, as evidenced by the noticeable share of productivity gains accrued to labour, which in turn provided the means of supporting mass consumption and mass production, the linchpin of the Fordist growth strategy. Other institutional forms, notably forms of competition, inter-state relations and money forms, were (partially) defined in a set of international agreements, most importantly those constitutive of the 'embedded liberalism' compromise reflected in the Bretton Woods settlement.

The relative stability of a growth model, for example, Fordism, does not imply a lack of conflicts, but rather that conflicts which invariably characterise the accumulation process and its socialisation are embedded, regularised and contained within the existing institutional forms. Secular crises—like that of the 1970s which prompted the emergence of the regulation approach itself—occur when existing institutional arrangements appear incapable of managing or containing social conflicts and the previous connection between the growth strategy and the overall mode of societalisation breaks down (Brand et al. 2008). Distinct from the actual crisis, the way in which the causes of and solutions to the crisis of Fordism were politically constructed as the neoliberal ideology was becoming ascendant from the 1970s onwards helped open the way to a new growth regime and mode of regulation, schematically referred to as post-Fordism. While debate continues on whether post-Fordism as a formation—that is, its growth logic and attendant sociopolitical institutional forms—is stable in advanced economies, it can be argued that although its concrete manifestation is mediated by different social, cultural and political structures and path dependencies prevailing in various countries, it also boasts characteristics that make it a distinct mode of growth and regulation. Perhaps the predominant feature of post-Fordism in frontier economies is that competition has displaced the wage–labour nexus as the defining institutional form (Petit 1999). Far from being a universal or natural law of markets, competition is instituted differently in different historical periods (Harvey 2002). Under post-Fordism, changes to the institutional form of competition have been accompanied by changes in other institutional forms. The wage–labour one, for instance, became the subject of major changes since early on, with the aim of reorganising the workforce along more flexible lines so as to support production and consumption patterns characterised by flexibility, differentiation and perpetual innovation in the context of a more open economy (Mouhoud 2006). Alongside changes in the forms of competition and wage relations, the structural form of money also changed considerably. Perhaps the most important of such changes, following the deregulation of financial markets from the 1980s onwards, is financialisation, a phenomenon of various manifestations, but which, for our purposes, we can define as the dominance of financial over productive capital and the institutionalisation of its value-set (e.g. short-term profitability, shareholder value) to the detriment of employment and wages.

Although institutional forms are always conceived broadly and never as simply top-down juridico-political regulation, the state retains a commanding role in the reconfiguration of these institutional forms, even if its own shape changes in the process (Cerny 1997; Petit 1999; Brand et al. 2008). Jessop's work on the shift from the Keynesian Welfare State towards the Schumpeterian Workfare State orientation captures both the role of the state in mediating changes to various institutional arrangements and the profound effect they, in turn, have on the internal structures of the state and on the system of states. Before turning our attention to some of the most important features of this shift in the system of states, its limitations for our present purposes need to be acknowledged. Being conducted at a rather high level of abstraction, work on the shift towards the competition state orientation is primarily useful in describing the trajectory of the state in advanced economies rather than in adequately explaining it (Hay 2004). However, as the focus here is on outlining the main tendencies and pressures emanating from changes in frontier economies, the abstraction and parsimony of this body of work can be seen as a boon. Besides, as and when these tendencies become relevant to the core argument that preoccupies us here in relation to India and Brazil, attempts are made to move from a descriptive to a more explanatory account.

The most important limitation, perhaps, is that this body of work has been exclusively concerned with the nature and trajectory of the state in advanced economies. The very different pasts, socio-economic conditions and modes of integration in the world market that characterise developing countries have meant that their states have taken many forms, none of which fitted comfortably within the Keynesian Welfare State form during Fordism, or could be said to fit the Schumpeterian Workfare form now. Nonetheless, as noted at the start, the responsibility for a country's economic performance post-WWII was everywhere placed on the shoulders of the state. Indeed, this was more so in developing countries, where, more often than not, developing and catching up with more advanced economies became the basis of state's legitimacy, as well as the justification for the radical, if uneven, social transformations deemed necessary to achieve this goal (Vielle 1988; Woo-Cumings 1999; Johnson 1999). Differences notwithstanding, state forms in the developing world—as elsewhere—are not only a domestic expression, but also shaped—in the light of their subordinate status, more here than elsewhere—by competition in the world economy (Cumings 1999). It is because of the usefulness in teasing out some of the main pressures the shift towards the Schumpeterian Workfare State orientation generates in the world economy that we turn our attention to this body of work. That is to say that neither the Indian nor the Brazilian state form fits into the ideal of the Schumpeterian Workfare State, but that this orientation has generated pressures in the world economy that push countries seeking to catch up to follow the lead by attempting the transformation of their own institutional formations. As will be seen in the forthcoming chapters, both the Indian and the Brazilian state have embraced in different ways the international competitiveness orientation.

The pressures emanating from changes in the frontier economies constitute an even more forceful 'whip of external necessity' for contenders now compared with earlier because, as noted, changes to forms of competition are central to the dynamics of institutional changes in post-Fordism. The shift towards the Schumpeterian Workfare or Competition State orientation, as conceptualised by Jessop (1993, 2002, 2000) and Cerny (1997), pivots on gaining and maintaining international competitiveness having become the key aim of state policies. Although the competition state comes in myriad forms, structural competitiveness is central to its successful performance. This is understood as the global efficiency and strength of national economic structures, which depend as much on economic factors as on social and institutional forms, because the latter can substantially stimulate or hamper an economy's productive and competitive thrust (Chesnais 1986). The irony of adopting the competitive state orientation on the part of key frontier states was that they were shifting their focus on making activities that contributed to national wealth more competitive in the global market while abandoning—to varying degrees—earlier tools of industrial policy and other economic instruments aimed at strengthening their domestic productive structures. As is well known, the real and constructed crisis of Fordism in the 1970s generated considerable concerns in the US and Europe about the perceived loss of international competitiveness, especially in light of the Japanese success. These concerns set in train a number of institutional reforms, in the US and UK most notably, that redrew the boundaries between the economic–non-economic and private–public spheres in favour of private market actors. As it would turn out, leaving it to private market actors to define their own financial, industrial and technological strategies in the context of increasingly liberalised and internationalised financial flows tended to subordinate such strategies to short-term and often purely financial goals, with important negative consequences in the real economy (Chesnais 1991).

Whatever the merits of the new competitiveness orientation, the instruments adopted led to a shift in state orientation away from developing a range of strategic economic activities within national borders for reasons of self-sufficiency, full employment and so on, towards interventions in pursuit of competitive advantage in the global marketplace. These interventions are strategically orientated towards promoting innovation-driven structural competitiveness that is understood to derive from a broad set of economic *and* non-economic factors. As will become more evident in the discussion on the knowledge economy, the excessive reliance of the competitiveness orientation on the non-economic partly manifests itself in the relocation of wealth generation in the limitless creativity of the human mind, a move which relies on ever-larger appropriations of society's comprehensive productive powers. In other words, the new orientation demands changes that are not limited to the economic, but necessarily engulf the social sphere. Because the technological and societal orders are mutually constituted (Wiebe et al. 1987), a strategic concern with promoting innovation implies a transformation of state–society relations, as the state becomes ever more involved in harnessing non-economic factors and non-commodity forms of social relations in the service of structural competitiveness (Brand et al. 2008; Jessop 2001).

The change in demands placed on the state becomes apparent when the ideal forms of Keynesian Welfare States and Schumpeterian Workfare States are contrasted. As Jessop (1993, 2002) puts it, the objectives of the former regarding economic and social reproduction were to promote full employment in a relatively closed national economy primarily through demand-side management and to generalise norms of mass consumption through welfare rights and collective consumption. The distinctive economic and social objectives of the latter are the promotion of (product, process and organisational) innovation, the enhancement of structural competitiveness in a relatively open economy mainly through supply-side interventions and the subordination of social policy to the demands of labour market flexibility, cost-saving concerns and competitiveness overall (ibid.). This shift in orientation, and the concomitant changes in social institutional forms that mediate it, was not politically unstoppable, nor did it occur everywhere. On the contrary, involving contestation, improvisation and experimentation, these tendencies found different expressions in advanced countries, depending on the institutional peculiarities and complementarities that accounted for success in some settings and difficulties in adapting to the new exigencies in others (Coriat and Schméder 2006).

As this schematic rendering makes obvious, the state needed to be 'brought in' only in certain scholarly debates, for in practice, it had never gone away, although its role was being transformed. Indeed, as it would turn out, in conditions of a more open world market, the state was to achieve more rather than less. For instance, because under conditions of openness and financial liberalisation what can counter the national economy's vulnerability to volatile financial flows is the competitive strength of the national economy, the state, as noted above, has become actively involved in the comprehensive restructuring and reorientation of the domestic non-/economic resources towards perpetual innovation and international competitiveness (Cerny 1997; Lipietz 2001; Brenner et al. 2008). In the meantime, its role as the guarantor of social order and cohesion within its borders has not only been retained, but made more complex by the multiple demands of competition and, in general, by the fragmentation and increased level of societal tensions that have accompanied the shift towards post-Fordism. Despite more complex demands placed upon it, the state's role and power are often said to have 'hollowed out'. Rather than reflecting the state's irreversible emaciation, the 'hollowing out' image better captures the formation and configuration of new social and political forms across different levels, as evidenced by the important role local, regional and international spaces have gained as new spaces of political and economic organisation and regularisation (Brenner et al. 2008). Because the drive towards competitiveness—differently constituted in time/place—is a structural feature of the modern state, this spatial reorganisation is not new, but reflects the particular way in which states' historical role of managing the relation between the national and the global spheres has been reconfigured in post-Fordism. During the latter, while it appears to have lost control over national economy as an object of economic management when compared with its post-WWII form, the state becomes actively involved in the process of internationalisation itself. Put differently, under conditions of increased capital mobility in a more open world economy, most states have moved away from viewing the national economic space as the best starting point for pursuing economic growth, towards managing its insertion into global circuits of capital through a combination of extraversion and penetration, aiming to secure a net benefit from such process of internationalisation (Jessop 2002).

Even if they have further undermined its economic autonomy, the state has itself been implicated in enabling and facilitating the transformations that have resulted in the national scale having lost the primacy in economic and political organisation. Of the newly empowered spaces, it is the international that is most relevant to our discussion. One of the main points that deserves underlining in this regard is that, while the nature of their engagement with the world economy varies and is determined in no small part by their position in the global hierarchy, all states are more or less compelled by the pressures and constrains generated by the 'international competitive treadmill' (Jessop 1997) to improve the structural competitiveness of their economies, that is, to become competition states of sorts. This is one of the ways in which we can speak of the competition state as an internationalised state form. Moreover, as this process has been taking place under conditions of liberalisation, financialisation and privatisation since the 1980s, all states—although, again, not equally—have become increasingly dependent on international capital and financial markets, the internalisation of which dependence further perpetuates their orientation as competition states (Brand et al. 2008; Brenner et al. 2008).

The intensification of competitive pressures variously bearing upon all states has also generated another dynamic at the international level that is worthy of attention. As states become increasingly concerned with their international competitiveness, policy regimes at the international level become strategically more significant to them (Jessop 2002, 2000). Although the reconfiguration of state forms in post-Fordism has meant that the state is not the only actor in these international regimes—as captured by the shift from government to governance—its control over the national territory still grants the state a central role in them. States competing with each other in an effort to expand their own benefits from internationalisation come together in these international policy regimes in an effort to shape them in ways that favour the interest of sectors that appear to be the most important to their competitiveness and economic growth (Görg and Brand 2000; Jessop 2000). Viewed from this perspective, the increased tensions and contradictions emanating from the totality of existing international regimes stem not so much from states' emaciated capabilities or asymmetries of information, but from the asymmetries of power between and within different policy regimes and between the groups of actors that participate in them (Brand et al. 2008).

The increased participation of so-called rising powers—including India and Brazil—in various international regimes can more fruitfully be interrogated not in terms of them accommodating to or challenging such regimes, but in terms of them seeking to enhance the competitiveness of their own economies. That both the Brazilian and the Indian state have internalised the logic of international competitiveness can be in no doubt. Mirroring concerns raised within policymaker circles in the US, Japan and a number of European countries about loss of competitiveness in world markets, the first civilian government in 1986 acknowledged that Brazil's new growth strategy had to be based on 'genuine' structural competitiveness (Leiva 2008; Pedersen 2008). Against the backdrop of the crisis of Fordism, the more immediate international sources of these pressures were different: a combination of geopolitical and economic pressures emanating from Cold War enemies (USSR) and allies (Japan) in the case of the US, and the international debt crisis that brought to an end its highly indebted growth strategy in 1982 in the case of Brazil. The socio-economic reforms adopted in pursuit of structural competitiveness also differed in nature, extent and speed. That the gap between the two has widened further has only increased Brazilian state's concern with improving its international competitiveness. A clear indicator of this orientation is the significant import placed on innovation-based growth and 'genuine competitiveness' in all its industrial policies since the late 1990s. In the case of India, similar international pressures, and particularly the rise of China, provoked an even more decisive embrace of the competitiveness logic on the part of the state. Having underpinned India's inward-looking growth strategy since independence, *swadeshi*—an idea of various interpretations, all of which insist upon the value of the local over the remote—was redefined during the 1990s to justify the new orientation towards competitiveness. *Swadeshi* now meant going out in the world and winning, because as Manmohan Singh put it in 1995: 'India's economic destiny is safe only when India knows how to ... compete against everyone in the world' (cited in Alamgir 2009: 263).

Given the central role granted to innovation in securing structural competitiveness, it is not surprising that the internalisation of the competitiveness logic should manifest itself in both countries in a pronounced attention to high-tech sectors, although low-value-added sectors are by no means dismissed. Seeking to take advantage of its vast natural resources, Brazil hopes to become a 'natural knowledge economy', whereas the Indian state has the goal of transforming India into one of the top five global knowledge powers by 2020 (GoI 2013). An orientation towards innovation-based competitiveness is by no means a Gerschenkronian institutional innovation on the part of India and Brazil, but mirrors closely the rise of the knowledge-based economy imaginary in the frontier economies under the broader flag of post-Fordism. The reasons for this are not to be found in the record of the knowledge economy itself. It is true that expectations and hopes surrounding the potential and promise—economic or otherwise—of new technologies always run well ahead of actual performance (Borup et al. 2006). But even with this caveat, for all the potential of the new technologies, such as ICT (information and communications technology), biotechnology and nanotechnology, which past and current defenders of the knowledge economy repeatedly invoke, the world economy only boasted an average annual growth rate of 3.2% in the 1980–2010 period as compared with 5% for 1950–1980 (IMF 2010; Palma 2012). The reasons for this performance are the subject of continuous debates, but it is difficult to escape the conclusion that it has much to do with the terms of the concomitant rise of financialisation and the knowledge economy in the post-Fordist period.

Be that as it may, the point remains that while the imaginary of a new knowledge economy based on new technologies is yet to generate the expected rates of growth, it has already generated significant socio-economic transformations across the world. One way of sketching these transformations is through following the patterns of the new international division of labour that has emerged during post-Fordism. The reconfiguration of the global economy geography has broadly followed two logics: the 'cognitive production logic', which favours spaces with a high concentration of intensive knowledge activities and capabilities, and the 'flexible Taylorian logic', which is the fate of regions/spaces whose advantage remains in low-cost labour (Mouhoud 2006). Put differently, the former spaces are strategically positioned to play the *technological* competitiveness game, whereas the latter the (less advantageous and rather more arduous) *price* competitiveness one. While this polarisation does not permit the making of simplistic explanations of the North–South kind, the case remains that around 95% of all research and development (R&D) is undertaken in wealthy regions of the global economy (Smith 2010). This means that, seen from the perspective of the national economy, the chances of gaining a competitive position in the knowledge economy are not very strong for most catch-up contenders, although niche successes can and have emerged within them. This said, and notwithstanding the hitherto meager growth rates and highly skewed polarisation patterns that have accompanied the rise of the knowledge economy imaginary, it is difficult to find instances of domestic or international policymakers not making reference, in one way or another, to the necessity of competing in the global knowledge economy today. The following section discusses what this implies.

2.2 The Rise of the Knowledge Economy

Since the 1990s, the particular economic imaginary of the knowledge economy has acquired a central role in the way the economy is conceptualised and discussed. Because discursivity and materiality constitute each other, it has also been central in guiding the adoption of specific practices that have contributed to the instantiation and materialisation of the knowledge economy in practice. Certainly, the emergence of new technologies associated with the so-called Digital Revolution—ICTs, biotechnologies and later nanotechnologies, among them—were already portending the rise of a 'new economy' as problems associated with the Fordist growth regime were politically constructed by various groups to constitute irredeemable crises during the 1970s. But even when radically disruptive—something that cannot be said confidently about the Digital Revolution—technological changes have never succeeded in single-handedly ushering in a new growth regime. A critically important role in this transformation is played by economic imaginaries adopted by specific groups that go on to translate them into policies that, in turn, transform existing economic and societal relations in ways that secure the basis for the new growth regime—and associated technologies—to consolidate and expand. The knowledge economy is one such economic imaginary that, against the background of a constructed and actual crisis of Atlantic Fordism and alternative strategies/imaginaries (e.g. Toyotism or Sonyism) of the 1970s and 1980s, emerged as the guiding master-narrative and, moreover, as a state project (Jessop 2005, 2010, 2000). It was a state project in the sense that this economic imaginary would go on to provide a more or less coherent direction for state policy in different fields/scales so as to achieve not only the necessary economic, but also social transformations on which the knowledge economy could eventually be based, built and secured (ibid.).

A number of ideas referring to the emergence of a 'new economy' had been circulating in certain academic circles since the early 1960s in the US. From the economic, futurologist, sociologist and management tributaries contributing to this debate, the work of Peter Drucker (*The Age of Discontinuity*, 1969) and Daniel Bell (*The Coming of the Post-Industrial* Age, 1973) were perhaps the most influential in conveying the idea that industrial societies were in the cusp of becoming, variously, knowledge economies, information societies or post-industrial societies (Kenway et al. 2006). Greater degrees of wealth would be generated as societies evolved towards post-industrial societies, as the sources of wealth shifted from the energy-intensive manipulation and making of materials and physical goods towards the manipulation of information and knowledge. Against a backdrop of growing concerns about the future of the US economy, post-industrial imaginaries of a radical restructuring away from a model of mass production/consumption that appeared to have reached its economic and environmental limits towards a knowledge economy based on the limitless creativity of the human mind had increasing appeal in US policy circles from the 1970s onwards (Cooper 2008; Hester 2016). Indeed, it could be said that the US state was the first to have adopted the imaginary of the knowledge economy. In the receptive mood occasioned by the crisis of Fordism, the influence of these ideas simultaneously reached the OECD (Organisation for Economic Co-operation and Development) that, initially adopting the idea of 'the information society', moved to generate, in its distinctive and seemingly apolitical fashion, indicators to measure and spread best practices during the 1970s (Godin 2006). By the mid-1990s, helped in no small measure by the contribution of new growth and evolutionary economic theories, the OECD's focus on information was replaced by that on the much larger category of knowledge. Its formal embrace of the knowledge economy agenda and continued work since then continues to help normalise and stabilise the knowledge economy imaginary worldwide. Indeed, so influential has this imaginary been that the existence of the knowledge economy is now beyond doubt, its exigencies necessitating incessant mobilisation from states, businesses, societies and individuals in all aspects of their activities.

The knowledge economy was thus not a project devised by an ambitious catch-up state. As will be seen in the forthcoming chapters, in embracing the knowledge economy imaginary, the Indian and Brazilian states were taking their cues from the US and other OECD countries that, enjoying a competitive advantage in the emerging technologies and related sectors, were the first to adopt it. The gradual, variegated but nonetheless steady embrace—by appeal or apparent necessity—of this imaginary by states seeking to catch up or at least avoid falling behind has helped consolidate the reach of this particular way of organising socio-economic activities well beyond its origins. This expansion and consolidation has gone hand in hand with the scope of state activities broadening rather than narrowing. The ubiquitous links made between the knowledge economy and international competitiveness in most official policy papers make evident that the primary task on the part of state actors is to maintain a truth regime that naturalises the knowledge economy and all its necessary forms and transformations (Jessop 2005). Even in its 'hollowed out' form, the state plays a crucial role in propagating a broader societal and political vision and, more generally, in maintaining some form of social cohesion in the face of continuous social conflicts and tensions. Again, legitimation on the basis of a broad and appealing vision is not specific to the knowledge economy, for successful growth strategies have always depended on a popular mobilising imaginary. The distinctive feature here is not novelty, but the manner in which such a vision has to correspond adequately to the transformations that are deemed necessary to create the conditions for it to succeed. As a growth strategy explicitly based on perpetual innovation and a rather broad set of resources on which innovation appears to depend, the knowledge economy requires legitimation for the deeper penetration of micro-social relations and the significant mobilisation of non-economic/non-material resources, for example, skill development, lifelong learning, flexibility, risk-taking, entrepreneurialism and so on. The significant changes needed in social and cultural practices to accommodate this orientation are multiple and obviously not managed or controlled by the state. Speaking of the knowledge economy as a state project does not mean that the state is omnipresent, but that it plays a critical role in giving certain form and direction to the transformations needed to sustain this particular socio-politico-economic formation.

The state's role along these lines becomes evident when probing the assertion that growth and competitiveness rely on the non-economic in more comprehensive ways than before. One way this can be achieved is by observing the reconfiguration of human labour—and life in its human and non-human forms—in relation to value generation. Compared with the nature of the economy when David Ricardo decisively relocated the source of all value on human labour, the materialisation of the knowledge economy has gone hand in hand with what can be referred to as the *real* (i.e. not just formal) subsumption of human labour in the service of value generation. This means that value is not strictly generated in the factory of old, but extensively in the 'social factory' (Smith 2007: 13). As the economy has 'softened'—that is, value is generated to a larger degree by non-material elements such as research, information, planning and design—not only one's direct labour, but the totality of the social individual becomes the foundation of wealth in the knowledge economy. This totality depends not only on education and various other publicly funded programmes, but crucially also on the broader and collective processes involved in the creation of social knowledge and its transmission and reproduction within and between generations (Morris-Suzuki 1986). In other words, human creativity and the social relations in which it is embedded and reproduced have now come to be considered and exploited as potentially limitless areas where value can be generated. Again, the state does not control these, but it plays a key role in creating, sanctioning and sustaining the conditions for their valorisation. The move towards the real subsumption of labour is everywhere a social and political phenomenon: for instance, the appropriation of the inventive capacities of employees—only a partial manifestation of the real subsumption of labour—has been enabled by a series of legal and institutional transformations sanctioned to various degrees by the state. These have included changes in the form of firms (e.g. from the large horizontally integrated form in the late nineteenth century to the small, highly specialised research company of the twenty-first century) and labour relations (e.g. the shift from 'shop rights' to 'work for hire', where ownership and right of use of invention/creation belong to the company), among others (Coriat and Weinstein 2009).

The deeper appropriation of society's comprehensive productive powers does not capture the full extent to which life itself has been subsumed in the service of generating economic value. It a broad sense, life—human and otherwise—has always been a source from which economic value has been extracted or generated (via human labour) in line with the institutional forms underpinning a particular socio-economic formation. In other words, that organic or inorganic material from nature has or is used to generate value has nothing to do with nature itself and everything to do with the prevalent socio-economic arrangement in a specific place and time that determines what constitutes economic value. An important recent juncture in this trajectory was the opening up of new spaces of valorisation in the multitudinous and limitless biological processes, human or otherwise, made possible by changes in life sciences from the 1970s onwards (Cooper 2008; Rajan 2012). Compared with the excessive reliance on inorganic nature (e.g. coal, oil) in earlier phases, the emergence of new commodities ranging from biodiversity credits to genes, human tissues, genetically modified (GM) seeds and other subatomic commodities in our time signals something different: the real and not just the 'formal' subsumption of nature in the service of value generation (Smith 2007).

The real subsumption of nature is clearer in the recent and rather fashionable imaginary of the 'bioeconomy', a system where economic value is generated through the application of biotechnologies on life in its multiple forms. Bioeconomy describes less an existing socio-economic formation than the political and moral leadership necessary to instantiate, normalise and legitimise the real subsumption of labour and nature, and the growth model based on it. Like the imaginary of 'knowledge economy', of which it is part, 'bioeconomy' is an economic imaginary with powerful performative and mobilising effects. Indeed, when the OECD turned its attention to bioeconomy in its 2006 International Futures Programme—following developments in the US which culminated with life sciences replacing chemical and atomic research as national priorities in the 1990s (Hester 2016)—it made no secret of its aim to design a 'bioeconomy policy agenda' governments could adopt to simultaneously address social, environmental and competitiveness issues (Goven and Pavone 2015). While the causes of these issues are never explored, the comprehensive social and economic transformations required to capture 'the latent value incumbent in biological products and processes' (OECD 2006: 3) are justified simply by the promise bioeconomy appears to hold in resolving them.

As will become clearer from our discussion on the emergence of biotechnology later on, the conditions for the commodification of (certain) biological resources such as genes or DNA sequences had to be established first in order for a market in biological commodities to emerge, function and expand. The institutional transformations needed to create and stabilise the new areas of accumulation subsumed under 'bioeconomy' involve a number of actors, and the state, again, plays a central role in their shape and orientation, if only because it still remains one of the main arenas where social conflict plays out. It is in this respect that the knowledge economy and the bioeconomy imaginary have been particularly significant in orientating state action towards facilitating the real subsumption of labour and life/nature in the service of the new growth strategy. Such subsumption does not imply the commodification of everything, but rather points to the fact that the modalities of state intervention in the economic and the social sphere change in line with this orientation.

The complex and often contradictory nature of these interventions is visible, for instance, in states being simultaneously relied upon to invest in R&D, education, research infrastructure and other elements contributing to the long-term generation of common and social knowledge on the one hand, and in creating and sustaining the conditions for subordinating these elements to short-term economic calculations on the other (Smith 2004; Zeller 2007; Jessop 2000). Managing this process, as well as the social tensions that emanate from it, also falls within the state's purview. From the complex list of often contradictory tasks facing the competition state today, of particular relevance for our purposes is the role it is called to play in creating and maintaining the necessary conditions for the production and circulation of knowledge as a commodity and as capital in the new knowledge economy. Although it has taken various forms in practice, few states can afford to ignore this role altogether, now that the generation and exploitation of knowledge—the 'new capital'—appears to have become central to wealth creation and to the broader aim of gaining and maintaining a competitive advantage in the global economy. Intellectual property titles (IPs) are the main means that enable the circulation of knowledge as a commodity and as capital.

2.3 Intellectual Property Titles and Knowledge as Commodity and Capital

To argue that in post-Fordism the terms under which knowledge is produced, used and appropriated have become crucial to wealth creation and, consequently, a focal point where social and political conflicts converge is not to underestimate the role knowledge has played in previous socio-economic formations. Knowledge has always been part of social and economic activity, even if its role has perhaps been more visible during major shifts associated with long waves of technological innovation

(Jessop 2000). Because knowledge and innovation are a fundamental part of human history, the struggle for control over them is not a post-Fordist but rather a much older phenomenon that has taken different forms through time/space. Although less complex than the contemporary competition states, the (European) mercantilist states, for instance, were keenly aware of the importance of knowledge and technological innovation for the growth of domestic industries, which is why they engaged systematically in industrial espionage and enacted various policies designed towards attracting and retaining skilled labour, then central to technological development (Harris 1998). During the heydays of Atlantic Fordism, neither a reduction of states' long-standing concern with their competitive position in the global economy, nor a reduction of their role in promoting the development of innovative capacities as central to their growth strategies can be said to have occurred. Indeed, technological innovation as a source of economic growth and competitive strength remained important to the Keynesian Welfare State, even though the terms under which these aims were pursued were different. This was visible, for instance, in the US and European national innovation policies being driven by knowledge production and technological development promoted primarily through massive public investment in sciences, higher education and intercompany R&D cooperation, and particularly important in the US, through the national security state (Borrás 2008; Weiss 2014).

Focusing on knowledge in the context of the current growth regime is thus not justified on account of other economic systems not having been knowledge based, but rather on account of the fact that the conditions for the generation and use of knowledge have been transformed. Whereas the Keynesian Welfare State orientation emphasised big science, big scale and productivity growth, the contemporary competition state is orientated towards perpetual innovation and structural competitiveness, with a much greater emphasis on self-reflexive management, learning and flexibility on the part of individuals, firms and institutions alike (Jessop 1993). Perhaps the most noticeable marker of our times is that the transformation of knowledge into a strategic economic good and a commodity has gone furthest than in any earlier socio-economic formation. The conditions under which the transformation of knowledge into a commodity has occurred are neither natural nor inescapable. Contrary to neoclassical economic accounts that see advances in knowledge and technology as exogenous to economy and society—as 'manna from heaven' (Smith 2010)—the conditions under which knowledge is generated and appropriated as commodity are determined through social and political conflicts. Often driven by those groups with most to gain from the commodification of knowledge, the outcome of these struggles could not have been secured and regularised were the state's juridico-political powers not used towards this end.

No better example of the crucial role the state has played in the commodification of knowledge can be found than its sanctioning of IP titles. A shorthand for a disparate set of legal entitlements, IP titles confer exclusive (temporary) rights for the exploitation and commercialisation of intangible assets to specific groups, and together constitute a framework that governs the terms of access, exploitation and circulation of knowledge and information.[2] By virtue of determining who can own, control and make use of what type of knowledge and who cannot, the institution of IP is a fundamentally political and contested one. IP titles should be seen not as mere technicalities filed away in legal texts, but, insofar as they grant control over economically valuable processes and resources to their owners and deny others the capacity to use them, as dominant forms of social and economic power (Drahos 1996). For this reason, they are better seen as a concrete and particular manifestation of the continuous process of primitive accumulation—as a 'second enclosure movement' (Boyle 2003)—that through disembedding knowledge from its collective roots provides the foundation on which a market for knowledge can be built. This process is necessary because (exchange) value cannot be generated in the absence of markets and markets, in turn, cannot exist without scarcity. But ideas, knowledge and other products of human intellect are not scarce; on the contrary, knowledge is a social product and a public good, being both non-appropriable and non-rivalrous in nature (Hughes 1997; Stiglitz 1999). IP titles radically counter such nature: by virtue of excluding others and granting control over knowledge to private IP-holders alone, they deliberately create scarcity where none existed in order to enable the commodification and appropriation of otherwise plentiful, collective, non-rivalrous intellectual goods.

Obviously, only knowledge that could potentially generate economic value needs to be made scarce in this manner. Because economic value is a socially created phenomenon, the boundaries between the value and non-value (or economic and extra-economic) forms of knowledge are constantly redrawn in line with the prevailing growth strategy and its mode of socialisation. It is not a coincidence that the period post-WWII up until the mid-1970s characterised by massive public investment in scientific research and the national innovation system by the Keynesian Welfare State was also one dubbed as IP 'Dark Ages' (Silverstein 1991: 304), as the economic value of IP titles and patents in particular was significantly eroded compared with the interwar period. Likewise, that the materialisation of the knowledge economy since the 1970s has gone hand in hand with the global proliferation and strengthening of private IP should not be surprising, for the more wealth creation has become dependent on the generation and exploitation of knowledge, the more important the role IP plays in protecting the accumulated capabilities of certain groups in this process. The real subsumption of labour and life/nature in the service of value generation under the growth strategy of post-Fordism discussed earlier could not have occurred without the deeper commodification of knowledge in general and of previously non-economic knowledge forms in particular. This increased commodification has manifested itself not only in the noticeable strengthening of the terms of exclusivity granted to IP-holders and in their global geographical extension. It is also clearly observable in the expansion of IP titles over economically valuable processes/resources opened up by the new technologies—for example, computer codes, financial algorithms, business models, databases, human, animal and plant gene sequences and functions—and in the proliferation of fierce conflicts over them.

It has not been the imperative formulated by states themselves, but rather that formulated by IP-holders as a group—which, incidentally, does not necessarily overlap with that of inventors and creators—that has historically provided the thrust of IP law. The actual content of IP laws has varied in time and place, reflecting the compromises and outcomes of political struggles involving other groups. Ever since the first patent monopoly was issued in Venice in 1474, all the constitutive elements of the current IP regime—for example, knowledge as appropriable, IP as a natural right, territoriality of IP claims or IP as necessary to spur innovation—have been contested at one point or another (May and Sell 2006). During this long period, of the many different ways of dealing with knowledge generation and circulation that incessantly brought various actors in conflict with each other, a particular way emerged as 'objective' and 'rational'—that of granting control over knowledge to a specific and narrow group to the exclusion of others through the granting of IP privileges (later, rights). Perhaps the most relevant observation that can be drawn from the long and complex history of IP for our purposes is that its trajectory has been closely intertwined with the trajectory of the modern state itself. There would be no markets in knowledge had it not been for the state acting as guarantor, enforcer and legitimator of IP titles. For, as noted, it is through IP laws—law being the quintessential form of the power to produce social groups and other categories (Bourdieu 1987)—that knowledge becomes scarce and turned into a commodity and an asset. Because (IP) law is a complex machinery of control with powerful universalising effects and by virtue of it being guaranteed by the state—still the only actor with legitimate monopoly over violence—such imperative has gained an almost natural status, so much so that alternatives to this particular way of dealing with knowledge have become unthinkable. In other words, one of the most important roles the state continues to play is not only that of providing and safeguarding the necessary conditions for the transformation of social/collective knowledge into a private commodity form through IP titles, but also that of maintaining a truth regime within which such transformation is deemed as legitimate and natural. This truth regime has now become an intrinsic part of the knowledge economy as an economic imaginary and state project.

That such role is not new but has been central to earlier state forms can be illustrated by pointing briefly to the particular way in which states' preoccupation with their competitive position in the world market and with sectors deemed key to their economic growth generated radical changes to IP laws during the Second Industrial Revolution. Of the important changes ushered in socio-economic institutional forms during this period, the most revealing is the rise of a new corporate model in Germany and the US simultaneously—the two most ambitious catch-up countries at the time—that was predicated as much on industrial research, production and marketing as it was on the management of IP and finance capital (May and Sell 2006). In light of the high-growth prospects of sectors where this corporate model initially took hold—the electric and chemical ones—the US and the German state responded relatively rapidly (in 1871 and 1877, respectively) to corporate demands to amend patent laws so as to grant them the right to own IP, which up to then was granted only to individuals (Fisk 2003). This marks the moment in time when corporations, having already been recognised as a singular personality in law, were granted IP rights for the work created collectively by their employees, thus opening the way for the real subsumption of labour that would intensify with the onset of the knowledge economy by the end of the next century.

This momentous change was part of other changes to IP occurring at the same time: preoccupied as they were with increased international competition, key states more or less accepted wholesale the solution to 'across-the-border' patent protection that business actors had developed over the ten years leading to the 1883 Paris Convention on patents (Porter 1999).[3] This convention also marks the point in time when an international patent regime was institutionalised for the first time in history. The outcomes of these radical changes were many and are still playing out today. Of particular interest is the dramatic shift in patent ownership and use: from the beginning of the twentieth century, the vast majority of patents have been owned by big business that—with the partial exception of patent's 'Dark Ages'—have routinely used them as strategic tools to control markets, limit competition and extend technological rents (Drahos and Braithwaite 2002). The disastrous effects of such use took little time to emerge. Using the thick web of legal protection afforded by IP titles as a disguise for market-sharing, price and production arrangements between themselves, firms in many industries coalesced around 'knowledge cartels' of unprecedented complexity that came to control an ever-larger share of international trade flows at the beginning of the new century, peaking in the interwar period (Porter 1999).

What this brief historical example illustrates is not only the fundamental role of the state in creating and safeguarding the conditions for the commodification of knowledge via IP in pursuit of generating value and economic growth, but also the manner in which this role is shaped by the terms of international competition. The next historical chapter of this story was subsequently opened by the US state zealously pursuing anti-cartel policies post-WWII, less because of the association of German and Japanese militarism with the 'knowledge cartels' of the past and rather more driven by the aim of tilting the playing field in favour of the new American corporate model (Porter 1999: 270). This was the very same model that, taking hold initially in the automobile industry, would come to epitomise the whole growth regime of the post-WWII period. But, once again, it was the competitive pressures stemming from Europe and East Asia that caused the US state to reverse its anti-cartel stance at the peak of the crisis during the 1970s. As a result of this policy reversal, a proliferation of joint ventures, market-sharing agreements and equity arrangements have ensued from the 1970s to this day, sharing a similar logic, although not form, to the cartels of the early twentieth century. At the same time, as the new growth logic shifted to the deeper exploitation of knowledge and the knowledge economy imaginary took hold, the US state first, and other frontier states soon after, moved to bring IP 'Dark Ages' to an end, a move that would culminate with the institutionalisation of a global IP regime by the mid-1990s. The competition states of this period, not unlike their forerunners a century earlier, once again took the preferred solution to global IP protection developed by high-tech sectors and, through the 1994 WTO (World Trade Organization) TRIPS Agreement,[4] institutionalised it in global public law (Sell 2003). The material changes brought about by these two trends—the relaxation of anti-cartel regulation and the strengthening of IP protection globally—can be observed in the global concentration of capital in life science industries, perhaps the most emblematic materialisation of the trend towards the real subsumption of life/nature in post-Fordism. Around 44% of the global pharmaceutical market, 50% of the global seed market and over 80% of the pesticide one is controlled by the top ten companies in each respective sector (Kesič 2011: 218; Brand et al. 2008: 19). All rely extensively on using IP as a strategic tool to control and expand their markets, limit competition and extend the technological rents that IPs afford them.

Observing these continuities is as important as noting the departures occasioned concomitantly by the internationalisation of the competition state orientation and the emergence of the knowledge economy as a state project. As the previous discussion has highlighted, starting in the US, changes in frontier economies following the real and perceived crisis of Fordism have led to an intensification of international competition pressures under conditions of a more open economy and liberalised financial flows. States everywhere have more or less shifted their orientation towards enhancing the structural competitiveness of their economies in the global market, an orientation that, in contrast to earlier periods, demands a much more extensive and intensive mobilisation of economic and non-economic resources in the service of perpetual technological, organisational and institutional innovation. Innovation here is not an end in itself, nor is it necessarily directed towards meeting social needs or countering environmental degradation. Rather, it is the crucial means through which economic value can continue to be generated by opening up new areas of accumulation and transcending the limits that are placed on its path as a result of constant (and historically specific) social, economic and political conflict. If knowledge and innovation in various socio-economic formations were to be placed in a commodification–decommodification continuum, the post-Fordist formation would undoubtedly fall closer to the commodification pole. What sets the contemporary knowledge economy apart from previous formations is precisely the phenomenon of knowledge having become a fictitious commodity in the proper Polanyian sense, the most concrete manifestation of which is the emergence of firms specialising in the production of knowledge/information for sale in the market, a phenomenon with few historical precedents (Morris-Suzuki 1986; Coriat and Weinstein 2009). Knowledge now also functions as a fictitious asset/capital (Jessop 2007; Zeller 2007; Birch and Tyfield 2012). As will be discussed in the next chapter, knowledge functions as capital because regular income streams or 'rents' from knowledge are guaranteed by stronger IP titles, which, moreover, can also be sold and bought independently in the market. Intangible assets such as IP titles, alongside other intangibles such as R&D, digital platforms, data and human capital, today constitute a substantial part of a company's market value: the market value of intangible assets was estimated to have exceeded that of tangible assets in the US in 2006, accounting for over 90% of total corporate value for many life science companies (Bryan et al. 2017: 61).

As noted, the Keynesian state orientation and the attendant IP 'Dark Ages' have been gradually eclipsed by a growth strategy and orientation that, being more dependent on the generation and exploitation of knowledge than ever before, requires the state to play a more complex role in facilitating it. One crucial aspect of this role relates to IP, for the more value generation has come to depend on the appropriation of social/collective knowledge, the more important IP has become as the main tool for protecting the capabilities of private (predominantly large) firms to benefit from such appropriation. It is precisely for this reason that the US state, joined later by some European states and the Japanese state, acted on behalf of high-tech, brand-name goods and entertainment sectors in which they had a comparative advantage to institutionalise a global IP regime during the 1980s. As has been accounted in some detail by Susan Sell (2003) and many others, the story of the WTO TRIPS Agreement that ushered in this regime in 1995 was an impossible feat. This was not so much because TRIPS embodied the solution business actors had worked out among themselves to rectify what they considered weak global protection and enforcement of their main assets, IP titles. This had also happened with the institutionalisation of the international IP regime a century ago. But it was an impossible feat to have this hardly disguised strategy to use IP titles again as a means of entering lucrative markets, limiting competition and extracting rents turned into international public law at the WTO, an organisation formally committed to trade liberalisation. Whatever was said about the links between stronger global IP and free trade during the negotiation process, TRIPS was clearly a particular instance of US and EU's (implicit) industrial policies extending outside their borders so as to protect the competitive advantage of their most technologically advanced sectors. In the context of the knowledge economy, it was also a clear manifestation of frontier states becoming increasingly locked into the pursuit of technological rents on behalf of their most promising sectors in ever wider geographical areas and for longer time periods (Jessop 2000; Smith 2010).

For the first time in history, most states are now legally bound to respect the high IP protection standards mandated by TRIPS, regardless of the very different socio-economic conditions prevailing in them. Despite the fact that it was justified to developing countries—all of whom are net IP-importers—on account of 'better' IP protection and enforcement stimulating domestic R&D, increased investment flows, transfers of technology and, ultimately, economic growth, TRIPS and the global IP regime it institutionalised are evidently about safeguarding the competitive advantage of actors that pushed for it (May 2000). Being an agreement intended to protect the competitive position of frontier economies, TRIPS has obviously made the task of catching up in the knowledge economy a much more challenging task. This task had been difficult even under the international IP regime it replaced that was rather more permissive of national IP variations by comparison. Indeed, even during patent's 'Dark Ages', only a few states—most notably, Japan—managed the Gerschenkronian feat of adopting existing technological advances to skip various intermediate stages and leapfrog to frontier economy status. Many more found themselves caught in the dynamic structuralist scholars warned about: the adoption of technological advances developed elsewhere implied the de-articulation of pre-existing social values, which, as these orient the local creative processes, in turn led to a specific and monologic form of technological and economic progress largely benefitting those actors leading the technological frontier (Furtado 1997). With the onset of TRIPS, even this mode of technology diffusion and 'learning by doing', historically the main way in which frontier economies of today developed, has been curtailed and is not readily available to catch-up contenders. All the same, the gravitational pull of the knowledge economy project that emerged in frontier economies compels them to attempt the transformation of their socio-economic structures, if not to successfully catch up, at least to avoid falling further behind.

Had the formidable challenge of technological upgrading in the context of a rather more restrictive global IP regime been the only one, countries seeking to catch up in the twenty-first century would have had an easier job in their hands than is currently the case. The main reason for charting the main tendencies of post-Fordism and of the knowledge economy in this chapter was to point out the rather complex and often contradictory tasks states everywhere are under pressure to undertake, and the nature of the wider terrain in which they are forced to compete. It is in this terrain that catch-up contenders such as India and Brazil must succeed. The terrain's tendencies, their position in it and their domestic socio-economic make-up, coupled with the kind of Gerschenkronian institutional innovations these states may be able to design and see through, would no doubt affect their chances of success. Whatever the future holds, it is clear from the discussion so far that dealing with the institution of IP is not the only challenge these states face in successfully competing in the global knowledge economy. A focus on IP is necessarily limited but not unimportant. Because it contributes directly to creating the conditions under which knowledge is generated, appropriated and exploited for purposes of value generation, the institution of IP is central to the way knowledge economies operate today. Being an institution whose form has significant ramifications on the overall distribution of social wealth and power, it has also become an increasingly important node towards which social and political conflicts gravitate. As the analysis progresses, while the focus is on the role of the state in shaping this institutional form in the case of two key life science sectors, pharmaceuticals and agro-biotech, attention is also paid to other relevant institutional reforms undertaken with the aim of creating the necessary conditions for the economic imaginary of the knowledge economy to be built and consolidated in practice, as well as to the many social conflicts that gave shape to them in India and Brazil.

Notes

1. This can be seen, for instance, in the emergence of the so-called the *new G7*, consisting of the BRIC and MINT economies (the new G7 includes China, Brazil, India and Russia, plus Indonesia, Mexico and Turkey).
2. The term 'intellectual property' itself entered common parlance only in the early 1980s, as shorthand for a set of disparate legal entitlements, of which the most familiar are patents (protecting innovations), copyrights (protecting original forms of expression) and trademarks (protecting words and

symbols identifying goods and services). Generally speaking, a patent is a property right in inventions that grants its owner a limited monopoly on the manufacture, use or sale of the invention (which can be a process or a product). A copyright arises automatically upon the creation of original work (such as books, movies, paintings and computer programs). Copyright protects only the form or expression of ideas, not the underlying ideas themselves. A trademark is a phase, symbol or design used to identify the source of goods or services; trademark law primarily prevents competitors from infringing upon the trademark. Other forms of intellectual property titles include trade secrets, industrial designs, integrated circuits designs, geographical indicators and plant breeders rights.

3. The international agreements on patents and copyrights of the late nineteenth century were, respectively, the Paris Convention for the Protection of Industrial Property of 1883 and the Berne Convention for the Protection of Literary and Artistic Works of 1886. The international secretariat created to manage them, BIRPI, was superseded by the World Intellectual Property Organisation (WIPO) in 1967.
4. TRIPS stands for Trade-Related Aspects of Intellectual Property Rights, a euphemism to justify the inclusion of IPs under the WTO remit, since all IPs known and in use at the time were included in the agreement.

Bibliography

Alamgir, J. 2009. *India's Open-Economy Policy: Globalism, Rivalry, Continuity*. Abingdon: Routledge.

Becker, J., J. Jäger, B. Leubolt, and R. Weissenbacher. 2010. Peripheral Financialization and Vulnerability to Crisis: A Regulationist Perspective. *Competition and Change* 14 (3-4): 225–247.

Bergeron, S. 2004. *Fragments of Development: Nation, Gender, and the Space of Modernity*. Ann Arbor: University of Michigan Press.

Birch, K., and D. Tyfield. 2012. Theorizing the Bioeconomy: Biovalue, Biocapital, Bioeconomics or … What? *Science, Technology & Human Values* 38 (3): 299–327.

Borrás, S. 2008. Innovation Policy and Institutional Competitiveness in Europe and Denmark. In *Institutions and Politics*, ed. P. Nedergaard and J.L. Campbell, 53–72. Copenhagen: DJØF Publishing.

Borup, M., N. Brown, K. Konrad, and H. Van Lente. 2006. The Sociology of Expectations in Science and Technology. *Technology Analysis & Strategic Management* 18 (3–4): 285–298.

Bourdieu, P. 1987. The Force of Law: Towards a Sociology of the Juridical Field. *Hastings Law Journal* 38 (July): 805–853.

Boyer, R. 1990. *The Regulation School: A Critical Introduction*. New York: Columbia University Press.

Boyle, J. 2003. The Second Enclosure Movement and the Construction of the Public Domain. *Law and Contemporary Problems* 66: 33–74.

Brand, U., C. Görg, J. Hirsch, and M. Wissen. 2008. *Conflicts in Environmental Regulation and the Internationalisation of the State: Contested Terrains*. London: Routledge.

Brenner, N., B. Jessop, M. Jones, and G. MacLeod. 2008. State Space in Question. In *State/Space: A Reader*, ed. N. Brenner, B. Jessop, M. Jones, and G. MacLeod, 1–26. Hoboken, NJ: Wiley.

Bryan, D., M. Rafferty, and D. Wigan. 2017. Capital Unchained: Finance, Intangible Assets and the Double Life of Capital in the Offshore World. *Review of International Political Economy* 24 (1): 56–86.

Cerny, P.G. 1997. Paradoxes of the Competition State: The Dynamics of Political Globalization. *Government and Opposition* 32: 251–274.

Chesnais, F. 1986. Science, Technology and Competitiveness. *STI Review* 1 (Autumn): 85–129.

———. 1991. Technological Competitiveness Considered as a Form of Structural Competitiveness. In *Technology and National Competitiveness: Oligopoly, Technological Innovation, and International Competition*, ed. J. Niosi, 142–176. Montreal: McGill-Queen's University Press.

Chibber, V. 2005. Reviving the Developmental State? The Myth of the "National Bourgeoisie". *Socialist Register* 41: 226–246.

Cooper, M. 2008. *Life as Surplus: Biotechnology and Capitalism in the Neoliberal Era*. Seattle: University of Washington Press.

Coriat, B., and O. Weinstein. 2009. *Intellectual Property Right Regimes, Firms and the Commodification of Knowledge*. Comparative Research in Law & Political Economy, Research Paper No. 17/2009. Available at http://digitalcommons.osgoode.yorku.ca/clpe/130. Accessed 20 May 2016.

Coriat, P., and G. Schméder. 2006. Post-Fordism in a More Globalized Capitalism. In *The Hardship of Nations: Exploring the Paths of Modern Capitalism*, ed. P. Coriat, P. Petit, and G. Schméder, 311–340. Cheltenham: Edward Elgar.

Cumings, B. 1999. Webs with No Spiders, Spiders with No Webs: The Genealogy of the Developmental State. In *The Developmental State*, ed. M. Woo-Cumings. Ithaca, NY: Cornell University Press.

Drahos, P. 1996. *A Philosophy of Intellectual Property*. Aldershot: Ashgate.

Drahos, P., and J. Braithwaite. 2002. *Informational Feudalism: Who Owns the Knowledge Economy?* London: Earthscan.

Fisk, C. 2003. Authors at Work: The Origins of the Work-for-Hire Doctrine. *Yale Journal of Law and Humanities* 15: 1–69.

Furtado, C. 1997. *Obra Autobiográfica de Celso Furtado*. São Paolo: Paz e Terra.

Gerschenkron, A. 1962. *Economic Backwardness in Historical Perspective*. Cambridge: Belknap Press.

Godin, B. 2006. The Knowledge-Based Economy: Conceptual Framework or Buzzword? *Journal of Technology Transfer* 31: 17–30.

GoI. 2013. *Science*. New Delhi: Technology and Innovation Policy, Ministry of Science and Technology.

Görg, C., and U. Brand. 2000. Global Environmental Politics and Competition Between Nation-States: On the Regulation of Biological Diversity. *Review of International Political Economy* 7 (3): 371–398.

Goven, J., and V. Pavone. 2015. 'The Bioeconomy as Political Project: A Polanyian Analysis' *Science. Technology & Human Values* 40 (3): 302–337.

Harris, J.R. 1998. *Industrial Espionage and Technology Transfer: Britain and France in the Eighteenth Century*. Aldershot: Ashgate.

Harvey, M. 2002. Competition as Instituted Economic Process. In *Market Relations and the Competitive Process*, ed. S. Metcalfe and A. Warde. Manchester: Manchester University Press.

Hay, C. 2004. Re-Stating Politics, Re-Politicising the State. In *Restating the State?* ed. A. Gamble and T. Wright, 38–50. Oxford: Blackwell.

Hester, R.J. 2016. Biology as Opportunity: Hybrid Rule from a Molecular View. In *Hybrid Rule and State Formation: Public-Private Power in the 21st Century*, ed. S.L. Hurt and R.D. Lipschutz, 175–192. London: Routledge.

Hughes, J. 1997. The Philosophy of Intellectual Property. In *Intellectual Property: Moral, Legal and International Dilemmas*, ed. A. Moore. Oxford: Rowman & Littlefield Publishers.

IMF. 2010. World Economic Outlook. Available at https://www.imf.org/external/pubs/ft/weo/2010/01/. Accessed 10 March 2016.

Jessop, B. 1993. Towards a Schumpeterian Workfare State? Preliminary Remarks on Post-Fordist Political Economy. *Studies in Political Economy* 40: 7–40.

———. 1997. Capitalism and Its Future: Remarks on Regulation, Government and Governance. *Review of International Political Economy* 4 (3): 561–581.

———. 2000. The State and Contradictions of the Knowledge-Driven Economy. In *Knowledge, Space, Economy*, ed. J.R. Bryson, P.W. Daniels, N.D. Henry, and J. Pollard, 63–78. London: Routledge.

———. 2001. What Follows Fordism? On the Periodization of Capitalism and Its Regulation. In *Phases of Capitalist Development: Booms, Crises, and Globalization*, ed. R. Albritton, M. Itoh, R. Westra, and A. Zuege, 282–299. Basingstoke: Palgrave.

———. 2002. *The Future of the Capitalist State*. Cambridge: Polity.

———. 2005. Cultural Political Economy, the Knowledge Based Economy and the State. In *The Technological Economy*, ed. A. Barry and D. Slater, 142–164. London: Routledge.

———. 2007. Knowledge as Fictitious Commodity. In *Reading Polanyi for the 21st Century*, ed. A. Bugra. Basingstoke: Palgrave Macmillan.

———. 2010. The Knowledge Economy as a State Project. In *The Nation-State in Transformation*, ed. M. Böss, H. Bang, and J. Campbell, 110–129. Santa Barbara: Aarhus University Press.

Johnson, C. 1999. The Developmental State: Odyssey of a Concept. In *The Developmental State*, ed. M. Woo-Cumings, 32–60. New York: Cornell University Press.

Kenway, J., E. Bullen, J. Fahey, and S. Robb, eds. 2006. *Haunting the Knowledge Economy*. London: Routledge.

Kesič, D. 2011. Strategic Development Trends in the World Pharmaceutical Industry. *Managing Global Transitions* 9 (3): 207–223.

Leiva, F.I. 2008. *Latin American Neostructuralism: The Contradictions of Post-Neoliberal Development*. Minneapolis: University of Minnesota Press.

Lipietz, A. 2001. The Fortunes and Misfortunes of Post-Fordism. In *Phases of Capitalist Development: Booms, Crises and Globalizations*, ed. R.M. Albritton, R. Westra, and A. Zuege, 17–36. Basingstoke: Palgrave.

May, C. 2000. *A Global Political Economy of Intellectual Property Rights: The New Enclosures?* London: Routledge.

May, C., and S.K. Sell. 2006. *Intellectual Property Rights: A Critical History*. London: Lynne Reinner Publishers.

Morris-Suzuki, T. 1986. Capitalism in the Computer Age. *New Left Review* 160 (November–December): 81–91.

Mouhoud, E.M. 2006. The Economic Geography of Post Fordism: On Knowledge and Polarization. In *The Hardship of Nations: Exploring the Paths of Modern Capitalism*, ed. B. Coriat, P. Petit, and G. Schméder, 293–307. Cheltenham: Edward Elgar.

OECD. 2006. *Scoping Document: The Bioeconomy to 2030 – Designing a Policy Agenda*. Paris: OECD.

Palma, J.G. 2012. *Brazil's Recent Growth*. UNCTAD and South Centre Discussion Paper No. 3 (Geneva: UNCTAD).

Pedersen, J.D. 2008. *Globalization, Development and the State: The Performance of India and Brazil Since 1990*. Basingstoke: Palgrave Macmillan.

Petit, P. 1999. Structural Forms and Growth Regimes of the Post-Fordist Era. *Review of Social Economy* 57 (2): 220–243.

Porter, T. 1999. Hegemony and the Private Governance of International Industries. In *Private Authority and International Affairs*, ed. A.C. Culter, V. Haufler, and T. Porter. Albany: State University of New York Press.

Rajan, K. 2012. The Capitalization of Life and the Liveliness of Capital. In *Lively Capital: Biotechnologies, Ethics, and Governance in Global Markets*, ed. K. Rajan, 1–43. London: Duke University Press.

Sell, S. 2003. *Private Power, Public Law: The Globalization of Intellectual Property Rights*. Cambridge: Cambridge University Press.

Selwyn, B. 2011. Trotsky, Gerschenkron and the Political Economy of Late Capitalist Development. *Economy and Society* 40 (3): 421–450.

Silverstein, D. 1991. Patents, Science, and Innovation: Historical Linkages and Implications for Global Competitiveness. *Rutgers Computer and Technology Journal* 17 (2): 261–319.

Smith, N. 2007. Nature as Accumulation Strategy. *Socialist Register* 43: 1–21.

Smith, T. 2004. Technology and History in Capitalism: Marxian and Neo-Schumpeterian Perspectives. In *The Constitution of Capital: Essays on Volume One of Marx's Capital*, ed. R. Bellofiore and N. Taylor, 217–242. New York: Palgrave Macmillan.

———. 2010. Technological Dynamism and the Normative Justification of Global Capitalism. In *Political Economy and Global Capitalism: The 21st Century, Present and Future*, ed. R. Albritton, B. Jessop, and R. Westra, 25–41. London: Anthem Press.

Stiglitz, J. 1999. Knowledge as a Global Public Good. In *Global Public Goods: International Cooperation in the 21st Century*, ed. I. Kaul, I. Grunberg, and M. Stern, 308–325. New York: Oxford University Press.

Vielle, P. 1988. The State of the Periphery and Its Heritage. *Economy and Society* 17 (1): 52–89.

Weiss, L. 2014. *America Inc.* Ithaca, NY: Cornell University Press.

Wiebe, B., T. Hughes, and T. Pinch. 1987. *The Social Construction of Technological Systems.* Cambridge, MA: MIT Press.

Woo-Cumings, M. 1999. Chalmers Johnson and the Politics of Nationalism and Development. In *The Developmental State*, ed. M. Woo-Cumings, 1–31. London: Cornell University Press.

Zeller, C. 2007. From the Gene to the Globe: Extracting Rents Based on Intellectual Property Monopolies. *Review of International Political Economy* 15 (1): 86–115.

CHAPTER 3

Financialisation and the Emergence of Biotechnology

That financialisation and the rise of the knowledge economy occurred concomitantly was not merely a happenstance; on the contrary, these two most visible tendencies of post-Fordism are closely intertwined. The close connection between the two can be observed in the transformations that led to the institutionalisation of a global IP regime from the mid-1990s onwards, although its roots go back much earlier. Ever since private firms were granted the legal right to appropriate the fruits of intellectual and collective labour of their employees through changes to labour and IP laws in the late 1800s, IP titles have been used as strategic business tools to protect and extend firms' market position and the period during which surplus profits guaranteed by IP could be maintained. Although it confounds those who defend IP as the only means of spurring creativity and innovation, this phenomenon is in line with evidence suggesting that historically creative periods have often preceded rather than followed the appropriation of knowledge through IP (Daly and Cobb 1989). As discussed in the previous chapter, what is distinctive about the knowledge economy is that this socio-economic formation relies on a more extensive and intensive mobilisation of economic and non-economic forms of knowledge in the service of value generation. The more the generation of value has come to depend on the appropriation of social/collective knowledge, the more important IP titles have become to those who benefit from such appropriation. It is for this reason that high-tech sectors, initially

V. Muzaka, *Food, Health and the Knowledge Economy*, Building a Sustainable Political Economy: SPERI Research & Policy, DOI 10.1057/978-1-137-59306-1_3

in the US and later in Europe and Japan, mobilised from the late 1970s onwards to strengthen what they saw as a weak and inadequate international IP regime, an aim they achieved in the form of the global IP regime ushered in by the WTO TRIPS Agreement in 1995.

Being part of the edifice of international public law, TRIPS can be seen as an effort on the part of frontier states to protect and extend the rents IP titles offer to their most technologically advanced sectors in ever wider geographical areas and longer time periods. Such efforts have gone hand in hand and facilitated by the broader tendency towards rent-seeking that is characteristic of the process of financialisation taking place during the same period. Following the deregulation of financial markets initiated in the early 1980s in frontier economies, the logic of financialisation that has taken hold since then is one of shifting the generation of value away from productive activities towards value accrued from ownership of assets broadly defined, that is, towards rents accrued not only from financial assets but also from knowledge assets such as IP titles (Zeller 2007; Birch and Tyfield 2012). As noted at the start, IP markets have grown substantially since the 1980s, as have rents accrued from an ever-growing number of IP titles worldwide. These rents take different forms that are not limited to receipts of licensing arrangements but include, perhaps more importantly, income generated from 'playing the stock market', where the value of high-tech companies in general and life science companies in particular is routinely inflated and consists, for the most part, of intangible rather than tangible assets. That this rentier instinct among market players should co-exist with a pronounced focus on innovation along (nominally) Schumpeterian lines is not a fallacy of composition but reflects a key contradiction in post-Fordism: the (Schumpeterian) drive to appropriate surplus profit through perpetual innovation goes hand in hand with efforts to protect such profits from competition and extend them temporally and spatially through various means, including monopolies of the kind provided by IP titles (Zeller 2007; Smith 2010; Jessop 2000).

These tendencies of post-Fordism have not only been shaping the global terrain in which catch-up contenders must compete and succeed, but also their domestic socio-economic make-up. The extent to which this has occurred so far depends on many factors, not least their position in the global economy and the strategies followed by state and business actors within them. For instance, partly as a result of their different historical patterns of insertion into the global economy and partly as a result of the different ways in which the Brazilian and the Indian state have responded to the rise of the knowledge economy and financialisation, the Brazilian economy has become a much more important platform for financial valorisation than the Indian one, and both have become sources of technological rents through IP for predominantly foreign life science companies. These different trajectories will determine to a large extent their chances of becoming competitive knowledge economies, but it bears repeating that the dice are loaded against them and developing countries as a group more broadly. This is not only the case because mastering the core technologies and operating near the technological frontier—central to the ability to compete in the knowledge economy—have become harder, thanks to a more restrictive global IP regime set up to preserve rather than close the technological gap. It is also because many of them were weakened and thrown off course precisely as the rules of the new global economy were in the making during much of the 1980s and 1990s. This was due in large part to the debt crisis of the early 1980s that abruptly forced many of them to abandon a model of development predominantly based on cheap borrowing and public investment and adopt an export-led model with limited borrowing facilities (Coriat and Schméder 2006). Being part and parcel of the broader transformations of post-Fordism, and especially of financialisation that altered the balance between financial and productive capital in favour of the former, the debt crisis pushed many developing countries on their back foot. Of course, as discussed in the previous chapter, financialisation in a context of a more open economy made states everywhere more dependent on financial and capital markets, a dependency that exacerbated the competitive pressures brought to bear upon them. But these pressures were even more acute for most developing countries, now forced to adopt a restrictive and politically problematic financial discipline deemed necessary to attract investment flows, the main source of capital they could rely on in order to compete in the world market (Coriat and Schméder 2006; Brand et al. 2008).

Such abrupt change in their trajectory stalled some of their economic sectors and pushed others into a tailspin. Especially those countries that were most heavily impacted by neoliberal disciplining during the 1980s saw the unravelling of many of the economic successes they had hitherto achieved as a result of import substitution strategies that partially imitated the Fordist growth model. The integration of economies at different levels of development and technological sophistication has historically had as a first casualty the most advanced economic activity in the less advanced party (Reinert 2009). The restructuring of many developing countries' economies along neoliberal lines during the 1980s and 1990s proved no exception: the nascent high-tech sectors were typically the worst hit. In many cases, the result of such restructuring was not the 'creative destruction' of the Schumpeterian kind, but the uprooting of key components of existing national innovation systems and the interruption, if not demise, of domestic technological accumulation processes due to economic surplus being siphoned off to (often foreign) financial institutions, or due to key domestic firms being bought out/merged with foreign companies, or both (Chesnais 2004; Balbachevsky and Botelho 2011). It is worth underlining that these processes of de-articulation of previous domestic productive and innovation linkages were taking place precisely as frontier states were starting to aggressively protect the competitive advantage and strategic assets of their high-tech sectors.

One way of untangling some of these complex developments is that of tracing the consequences of the considerable rise of interest rates in 1980–1981 in the US, the so-called Volcker shock,[1] which would provoke the debt crisis and the economic plunge of many developing countries during that decade. Against the backdrop of the crisis of Fordism, the shift to monetarism in the US—of which the interest rate increase was part—did not affect its position as a frontier economy but in reality strengthened it. As will become clearer when discussing the emergence of biotechnology in the next section, high federal R&D levels were more or less maintained through the floating of massive issues of T-bonds and other forms of government debt at a time when most other states had to reduce theirs, often quite drastically. The aim of such spending was as much geopolitical as it was the strengthening of US technological base and competitiveness vis-à-vis other countries (Cimoli et al. 2008; Weiss 2014). Alongside the rise of the Washington Consensus and financial liberalisation measures, these developments had at least two pernicious effects on developing countries' chances of moving up the technological ladder. First, they (unevenly) brought about and legitimised the apparent necessity of state withdrawal from the economy in general and innovation-related investment in particular in many developing countries, but, importantly, not in the US. Second, they opened the way for the rise of international financial capital flows whose short-term and largely speculative/rentier nature was and remains completely at odds with the kind of financing required to build innovative and dynamic productive capabilities: stable and rather long-term investment and engagement with the technological and learning processes in the real economy.

The indisputable US leadership in biotechnology that was the result of some of these developments is outlined in the following section with two aims in view: first, to show that the emergence of this sector was less the outcome of the American entrepreneurial spirit supported by deep capital markets and rather more the work of the visible hands of the US state. Second, attention is paid to the particular form the biotech sector took in the US, especially to the rapid enclosure of the frontiers opened by the new technologies by private IP titles and finance. This particular form would go on to shape not only the rise of biotech sectors in other countries, but, as the applications of biotechnologies are wide, also the trajectory of cognate sectors such as pharmaceuticals and agro-business. Moreover, constituting one of the core technologies of the knowledge economy, the development of biotechnology in the US reveals many of its dynamics: the intensification of the subsumption of labour and nature in order to generate value, the use of IP to protect the capability of certain actors to appropriate such value, and the way in which high expectations about its imagined potential to resolve a wide array of problems legitimised measures that helped secure the basis on which this sector could consolidate and expand in practice. These dynamics and the pioneering institutional changes that undergirded them emerged in the US but were not limited to it; under heightened competitive pressures, other developed countries and catch-up contenders attempted with different speeds and levels of success the transformation of their own institutional arrangements in ways that would enable them to capture part of the economic value created by the opening of this new technological frontier. In the case of India and Brazil, it can be argued that reforms towards this end represent less radical institutional innovations of the Gerschenkronian type than (locally refracted) emulations of institutional changes that took place in the US during the 1980s. This is another reason why attention to the most important among these US institutional innovations is paid in the following section.

3.1 The Visible Hand at Work: The Development of Biotechnology in the US

The US Federal Reserve's shift towards a more aggressive monetary policy was neither an exclusively internal affair nor an isolated one. The significant rise in interest rates that contributed to the debt crisis that would engulf much of the developing world was not the only revealing sign of the changes that were in the making. Those on the watch would have noticed at least three other significant events unfolding in 1980 alone: the Supreme Court decision on *Diamond v. Chakrabarty* in June, the first-ever initial public offering of a biotech firm (Genentech) in October, and the Bayh–Dole Act in December. The Supreme Court *Diamond v. Chakrabarty* decision allowed for the first time a patent to be granted on a genetically engineered bacterium, a move that ushered in significant changes to patents in the US and, later, worldwide. First, like its decision that had expanded the patentable subject matter to include computer software in the early 1970s, the 1980 Supreme Court decision brought to an end the numerous preceding legal battles over the question of 'life-patents' by further expanding patent rights to another emergent area of the Digital Revolution: genes and living matter. This decision was followed not long after by the grant of the first (utility) patent to a genetically modified seed (the *Hibberd* case) in 1985 and the first patent on a genetically modified (GM) animal (the Harvard onco-mouse) in 1988 (Safrin 2004). Second, although patentability was justified partially on account of the human inventive labour that created a GM organism such as the oil-eating bacterium, the decision opened up the way for patents to be claimed on genes and DNA sequences that had not been modified, that is, on discoveries that up until that point in time were explicitly excluded from patentability. Third, and related, the monopoly effects of such extension of private property titles over knowledge that had hitherto belonged to the intellectual commons were exacerbated by the nature of these titles. The US Patent Office has since then explicitly recognised and granted 'life-patents' of broad scope that cover not only the application disclosed in the patent application, but also a wide range of possible *future* applications, thus legalising the pre-emptive use of patenting to exclude both existing and future competitors (Coriat et al. 2003; Zeller 2007). Patent's 'Dark Ages' were over.

Following on the heels of the landmark 1980 Supreme Court decision, the expansionary IP trend in the US was further institutionalised with the creation of the centralised and specialised Court of Appeals for the Federal Circuit (CAFC) in 1982. Replacing the multi-circuit, largely unpredictable and patent-hostile court system in existence during patent's 'Dark Ages', the CAFC greatly expanded the reach of IP by upholding most of the patent claims, awarding generous awards for damages, lowering standards for patentability and expanding the definition of patentable subject matter to new areas, in line with the demands of business sectors with high stakes in patent protection that mobilised to create it (Merges 2000;

Landes and Posner 2004). The number of patents granted by the US Patent Office expanded rapidly, but the rise was more pronounced in biotech patents, whose numbers rose by 15% a year between 1990 and 2000 (Zeller 2007: 94). The end of patent's 'Dark Ages' was accompanied by the simultaneous cooling of US state's enthusiasm for anti-trust measures and a complete reversal of its earlier suspicion of private *pre-competitive* collaborative arrangements between firms by the late 1970s (Porter 1999: 271–274). This shift, in turn, followed and further facilitated changes in the corporate model of that period marked by the vertical disintegration within large (Fordist) corporations and, importantly for our purposes, the rise of new, smaller firms, often in the form of highly specialised, high-tech and science-based firms, which, in life science sectors at least, would become the mainspring in the dynamics of innovation and the object of the new inter-company pre-competitive collaboration networks (Coriat and Weinstein 2009).

In the life science industries, the new specialised high-tech firms materialised in the hitherto unknown form of dedicated biotech companies whose assets were primarily and often exclusively based on scientific knowledge. The origin of these firms is to be found in the manner in which concerns about the US competitiveness and its 'innovation crisis' opened the way for the knowledge economy imaginary to take hold in the US. Of particular interest here is the manner in which concerns about loss of the US competitiveness was used to politically justify changes to the orientation of its science and technology (S&T) policy during the 1970s that would push many universities into becoming actively involved in re-establishing the US competitive and technological leadership position (Coriat et al. 2003; Berman 2014). The most important industry–university partnership to inaugurate Nixon's conscious shift away from the military–university partnerships that had prevailed in the US until the late 1960s was that between Monsanto—one of the US largest chemical companies at the time—and Harvard University in 1974 in pursuit of molecular biology's potential (Hurt 2016). Institutional innovations designed to enhance these kinds of partnerships soon followed: among them, the 1977 National Science Foundation SBIR (Small Business Innovation and Development) programme—currently one of the largest *public* investment programmes in the world—followed a year later by the Industry–University Cooperative Research Centres programme (Weiss 2014; Berman 2014). Concerns about the ownership of IP titles over research outcomes arising from federally funded programmes such as these were addressed decisively in the Bayh–Dole Act in December 1980. This landmark Act abolished the established 'open science' principle predominant during Fordism by formalising rules related to universities' right to privately own patents on publicly funded research outcomes, opening the way to the appropriation and commercialisation of hitherto freely available academic research. Following the US example, Brazil would later enact its own 'Bayh–Dole' Act (Innovation Law of 2004), whereas the fate of the Indian Utilisation of Public Funded Intellectual Property Bill of 2008, aiming for the same, is yet to be decided.

Having broken down the established boundaries between the 'republic of science' and the 'kingdom of technology', that is, between freely available publicly funded basic research to which biotechnology owes its emergence and applied technology often occurring within industry, the Bayh–Dole Act pushed many US universities into the patenting race. In the area of human biology alone, the number of patent applications by universities increased by 300% in the first five years following the Act (Drahos and Braithwaite 2002: 163). Reflecting their interpretation of the Act as a duty to commercialise and patent, the number of patents held by US universities between 1979 and 1997 rose nearly tenfold, compared with a twofold increase in patents overall (Jasanoff 2005). Once publicly funded research was enclosed within patents walls, many universities sold their patents to companies or set up joint ventures to exploit them. Perhaps the most visible transfer of high-profile scientists into entrepreneurial entities found expression in the spectacular boom in university 'spin-offs', which in life sciences often took the form of dedicated biotech firms (Coriat et al. 2003). One of the first and most successful of these firms, Genentech, was founded by Herbert Boyer (University of California), who—alongside Stanley Cohen (Stanford University) and building on the work of scientists too many to mention and decades of accumulated knowledge in numerous disciplines—was one of the firsts to transplant genes from one living organism to another (rDNA), now a basic technology of genetic engineering. When the *Chakrabarty* decision cleared the way to 'life-patenting', the US Patent Office turned its attention to the backlog that had been building up in this field, starting with Cohen/Boyer's patent application(s) on basic methods of gene splicing. That claims over these and other patents related to interferons[2] would take some time to be approved and become public did not come in the way of Genentech achieving what until then had simply been unimaginable: one of the largest initial public offerings against a background of financial gloom of a company that had no products or earnings but whose valuation, despite the absence of similar listed companies, reached the unexpectedly high US$530 million[3] mark at the end of the first day of furious trading on 14 October 1980 (Rasmussen 2014: 120). A new chapter in the history of markets for knowledge had opened and finance was its key protagonist.

Anticipation, expectations and imaginaries related to technological progress are a typical modern phenomenon, but one that is expressed differently in time and place. In line with the rise of the knowledge economy and the broader shift towards post-Fordism, for instance, the future promise and potentiality of S&T have become decidedly more significant and intense. Such expectations and promissory imaginaries are not only performative—mobilising, as they do, fundamentally necessary real-time activities that turn the future into the present—but also loaded with values of different kinds: social values linking various, asymmetrically positioned social groups and communities and, more importantly for our discussion, economic value (Brown 2003; Borup et al. 2006). That expectations are constitutive of economic value was amply demonstrated by Genentech's initial public offering; its surprisingly high valuation was in keeping with the hyperbolic expectations attached to biotechnology, notwithstanding the fact that no biotech product was anywhere close to entering the market then and only a handful of successful ones have materialised to date (Pisano 2006; Hopkins et al. 2007; Birch and Tyfield 2012). Of all the new frontier technologies, the future promise of biotech has perhaps been the most susceptible to hype: many biotech companies were more overvalued during the dot.com bubble than companies in other sectors, a trend that, after a downturn during the recent financial crisis, resumed in 2013 (Shubber and Bullock 2015). In Genentech's case, value was created not only in the absence of products and earnings, but also in the absence of property titles such as patents; the only significant 'assets' Genentech had in October 1980 were its high-profile scientists and the future promise of its product and patent pipelines.

Genentech's initial public offering represents not a unique and deviant case, but rather an early example of the particular way in which financial capital, emerging from the shackles of national regulation, came to colonise this emerging scientific and technological frontier. Value would be generated not only through the intensification of the appropriation of social labour in general and collective and publicly funded scientific knowledge in particular. What made biotech firms an especially attractive investment for financial capital since their early days was the more-or-less secure income stream from IP titles such as patents (Coriat and Weinstein 2009), that is, the rents from (intellectual) property titles that make the private appropriation of knowledge possible in the first place. Unsurprisingly, the market value of biotech and pharmaceutical firms invariably increases when a patent has been awarded (i.e. long before a product reaches the market), a trend which exacerbates their tendency to multiply their IP titles and to zealously protect and extend them worldwide (Zeller 2007; Birch and Tyfield 2012). Like other financial investments, this too leads to overvaluation of assets, bubbles and crashes, of which the dot.com is the most notable example. Worse still, it can also lead to a 'tragedy of anti-commons' effect, which, many argue, has already hindered further technological innovation in biotechnology (e.g. Heller and Eisenberg 1998; Pagano and Rossi 2009), another contradiction stemming from the manner in which biotechnology developed in post-Fordism.

The entry of finance as a significant player in markets for knowledge and biotech science was the unique outcome of institutional changes in the US, in which the state played a key role. While the 'Volcker shock' had the effect of draining many developing countries of financial flows and directing them into the US in search of profitable investment opportunities, the role finance would come to play in the trajectory of biotech could only come about as a result of financial liberalisation measures undertaken by the US state from the late 1970s onwards. Two such measures are of particular interest here: the relaxation of 'prudent man' rules in 1979 and the 'Alternative 2' NASDAQ[4] regulation in 1984. The 'prudent man' rules in the US Employee Retirement Income Security Act had severely limited the amount of money pension funds could allocate to high-risk assets; following changes to this Act in 1979 which explicitly allowed pension funds to invest up to 10% of their capital in high-risk assets and venture funds, pension fund investments in these markets grew dramatically. Compared with 1978, when pension funds accounted for only 15% of the total US$218 million invested in new venture capital funds that year, this share had increased to 46% in 1988, by far the largest share of a (significantly larger) total of US$3 billion in new venture funds that year (Gompers 1994: 13). The effects of 'pension fund capitalism' would eventually be felt in capital markets worldwide (Clark 2000), not least because US immediate competitors and catch-up contenders such as India and Brazil sought to emulate these practices by pushing pension funds and other institutional investors to participate more actively in venture capital markets.

It is nevertheless worthy of note that their early participation in US knowledge and biotech science markets were amplified by the so-called Alternative 2 regulation of the US Security Exchange Commission in 1984, which authorised the listing of loss-making firms such as the dedicated biotech firms on the First Market of the national NASDAQ market, that is, on the most liquid and attractive NASDAQ market (Coriat and Weinstein 2009). Again, intent on deepening venture capital markets, other competitors, such as the UK and Germany, would attempt their own version of stock exchanges for high-risk companies (Alternative Investment Market and *Neuer Markt*,[5] respectively). As discussed, the value of these and other life science firms was and remains predominantly and often exclusively based on knowledge assets (IP titles) and, as in all asset markets, is highly sensitive to expectations of future returns on them. Any risks of anti-fraud litigation that could have emerged by the speculative inflation of promissory futures on the part of these kinds of firms were eliminated in 1995; reversing provisions enacted in the wake of the Great Depression, the Security Exchange Commission changed its safe harbour provisions for forward-looking statements allowing firms to make speculative statements, thus making hype and speculation not only an essential but also a legally mandated element of valorisation in knowledge markets (Brown 2003; Fortun 2012).

While conditions were thus being created for there to be a market of biotech science/knowledge and for financial investors to participate in it, the American public had more or less bought into the promise of biotech as a technology that would turn DNA into gold and provide the much-needed elixir to America's 'senescent' industries (Rasmussen 2014; Hester 2016). Such conviction could not have been based on the visible economic impact of scientific breakthroughs in molecular biology—there were many of the latter and none of the former—but rather resulted from a particular economic imaginary that was being propagated at the time. Amidst growing concerns that depicted the US economy being caught in the tightening grip of stagnation and loss of international competitiveness throughout the 1970s, a way out emerged in the form of the knowledge economy imaginary, which, springing out of various intellectual fountainheads, was to become a state project of the political right when Reagan came to power (Cooper 2008; Block 2011; Hester 2016). The foundations for the transformative changes specific to life science research had in fact been laid during the late 1960s and early 1970s, when, against a background of hostile relations between the Nixon administration and the scientific community, conscious efforts were made by the former to disassociate advances in molecular biology from their military past and link them to peace, economic growth and, importantly, health[6] (Hurt 2016). By the end of the 1970s, an abundance of private- and government-funded reports existed that linked the domestic sources of the economic malaise and its remedy, among others, to changes in S&T policy, funding and the IP regime (Berman 2014). In addition to the institutional changes highlighted above, the knowledge economy imaginary contributed to the enactment of a series of reforms by the new Reagan administration, the most immediate outcome of which was the dramatic increase in federal support for life science industries, which, replacing chemistry and atomic energy, would become the most heavily funded area of basic research and the primary national research focus (Cooper 2008; Hester 2016).

3.2 Reactions Outside the US

The oft-told story of biotech exploding into the scene, thanks to liberated financial flows and neoliberalism unleashing the entrepreneurial spirit does not stand up to scrutiny. Over the years, the US state had not only been busy creating the conditions for the emergence of a commercial biotech sector, but also funded it; indeed, the public hands of the US state are particularly visible in the financing of life science R&D. Drawing on a wide spectrum of scientific disciplines and on the slow accumulation of scientific knowledge therein, biotechnology owes its emergence to the massive public spending for basic scientific research in general and, especially important in the case of the US, for health-related molecular biology in particular (Orsenigo 1989). Venture capital typically entered the scene 15 years after considerable public investment had generated outcomes with promising commercial applications. This remains the case today, for even as total US federal funding—having accounted for more than half of national R&D—started to fall below industry's share from 1980 onwards (Weiss 2014: 99), massive federal funds continued to be channelled towards life science research. The US National Institutes of Health (NIH) alone, a federal agency of the US Department of Health, had contributed no less than US$790 billion (in 2011 US$) to molecular biology and health research in the US in the 1938–2011 period and around US$ 300 billion in the last decade (Mazzucato 2015), amounts that were simply not available to the immediate competitors of the US. By way of example, the public expenditure for life science research in the UK,

France and Canada together was only around 15% that of the US during the 1970s and 1980s (Borrus 2002: 16). By the mid-1990s, a European Commission concerned about the seemingly widening technological gap in biotechnology noted with dismay that public and private R&D investment in the sector were still over three to four times higher in the US than in the EU combined[7] (EC 1996).

These figures, however, indicated neither a lack of understanding of biotechnology's potential nor of attention to it in Europe. Key European states, for example, Germany, UK and France, acted with unprecedented urgency on the belief that biotechnology was a strategic technology potentially capable of impacting upon the performance of many economically important sectors, thus making the rapid closure of the technological gap vis-à-vis the US a key priority if negative long-term consequences in their international competitiveness were to be avoided (Orsenigo 1989). In all these three countries—incidentally, where publically funded research had since the 1950s been generating the startling discoveries that would provide the basis for the US biotech sector's success later on—it was the state that moved into action, without any urgent pressure coming from established cognate sectors such as chemicals or pharmaceuticals. As a matter of fact, some of the large companies that dominated these sectors were resistant to changes biotechnology posed to their established mode of operation (Orsenigo 1989; Cooke 2001). In Germany, since the 1970s and especially since concerns about Germany's 'poor performance' in IT and biotech compared with the US made promoting innovation a rare unifying goal of both left and right parties from the 1980s onwards, investments in biotech have been largely financed by the state (Cooke 2001). The most successful federally funded programmes were BioRegio and BioProfile in the 1990s, rolled out to achieve Germany's aim of becoming Europe's leader in biotech by the turn of the new century, followed up by the BioFutures programme in the 2000s, which, like the SBIR programme in the US, provided important sources of public funding for private biotech start-ups (Casper 2007).

The grant of the Cohen–Boyer patent in the US in 1980 prompted the Japanese government to take systematic action on biotech; taking the leading role, MITI[8] policies were directed explicitly towards the development and diffusion of biotechnology within the existing industrial system, with companies participating directly in public research programmes and the establishment of long-term goals (Orsenigo 1989). In keeping with its dirigiste orientation, the French state imposed its large-scale programme of industrial biotech innovation to a somewhat reluctant industry, although its great ambitions were deeply affected by the cuts in public expenditure during the late 1980s (Orsenigo 1989). In the UK, funding cuts for university research—the mainspring of important biotech discoveries such as that of the DNA structure in the 1950s and of monoclonal antibodies[9] in the 1970s—were so severe during the Thatcher years that by the mid-1990s, the UK was trailing behind not only the US, but also Japan, Germany and France in terms of the percentage of GDP spent on science (Jasanoff 2005). Earlier on, concerns about the rapid commercialisation of biotech in the US and fears that the UK was lagging behind had nevertheless been intense enough to prompt the government to establish Britain's first biotech company, Celltech, in 1980 (Senker 1996; Cooke 2001). Beyond this step, government funding remained relatively insubstantial and efforts focused instead on stimulating the formation of a venture capital market of the US kind to support private biotech start-ups (Orsenigo 1989; Jasanoff 2005). These efforts were followed by pressures on pension funds to increase their rate of venture capital investment, boosted by measures that saw the London Stock Exchange break NASDAQ's monopoly on high-tech/high-risk companies' listing in 1993 (Senker 1996; Cooke 2001).

Not only immediate competitors, but ambitious catch-up contenders were also alive to the potentialities of biotechnology. Like the Fordist state in the advanced economies, the state in India and Brazil had also supported traditional biology early on and, compelled by biotechnology developments in the US in particular and other advanced countries, it also took systematic measures to support the new technology from the 1980s onwards. But, as will be seen, the difficulties were enormous, not least because the considerable amounts of investment required were hard to muster. Only China—having embarked on the Four Modernisations strategy and the National High-Tech Programme, which was headed by biotechnology as the priority area at a similar time—appears to have managed the feat: the double helix became the symbol of China's high-tech drive and continues to gobble up considerable chunks of its total S&T investment, around 20% according to some estimates (Smith 2000; Chen et al. 2007). Even so, it is difficult not to conclude that although many developed and developing countries responded with unprecedented urgency to the pressure of building or supporting their nascent biotech sectors and had, by the 1980s, policies in place aimed at competing in this new technological frontier, none had the deep pockets of the

US state. Their (uneven) embrace of monetarism emerging from the US from the early 1980s onwards—with the notable exception of China—brought to bear upon them many of its exacting demands but few of its anticipated benefits, at least as far as strengthening their emerging life science sectors was concerned. The structural inequalities between a country whose national currency is also the international currency and the rest were glaring: while the rest of the world was subjected to a more or less rigid monetary discipline, the US could dramatically increase its support for life sciences, thanks to its unique position as the focal point of world financial flows and as the world's largest debtor (Cooper 2008; Block 2011).

Notes

1. Paul Volcker was the Chairman of US Federal Reserve, 1979–1987.
2. Interferons as a class of biological agents were known since the 1950s, but their generation changed with the emergence of rDNA techniques.
3. Around 2015 US$1.4 billion.
4. NASDAQ stands for the National Association of Securities Dealers Automated Quotations, which was founded in 1971. It is the second-largest exchange market in the world (by market capitalisation) after the New York Stock Exchange.
5. The Alternative Investment Market (AIM) was created in 1995 and is part of the London Stock Exchange, whereas the *Neuer Markt* was established as a new stock exchange with substantially less burdensome listing requirements than the existing blue-chip Frankfurt Stock Exchange in 1997. However, the dot.com crisis wiped an astonishing 90% of its value and the market was closed in 2003, with surviving companies transferred to the Frankfurt Stock Exchange.
6. Reference is made here to Nixon's 'War on Cancer' in particular and the fundamental contribution of the National Institutes of Health (NIH)—a federal agency of the US Department of Health and Human Services—to the accumulation of scientific and technological capabilities in the field of molecular biology and health in general. It is in part due to the NIH's role that the US has, above all, a competitive advantage in health biotech (as compared with, say, agriculture).
7. Definitional and accounting problems mean that comparisons across countries can only be approximate.
8. MITI stands for the Ministry of International Trade and Industry, one of the most powerful agencies of the Japanese state widely credited with Japanese economic success post-WWII (see Johnson 1982).

9. Monoclonal antibodies discovered by two Cambridge-based scientists (Milstein and Köhler) in the mid-1970s went on to have many applications in diagnostics as well as drug delivery and are generally held to have laid the foundations for applied biotechnology. They have certainly generated a multi-billion industry, but neither the researchers nor their funder, the UK Medical Research Council, got any returns as they chose not to patent the discovery but published the results as a letter in *Nature*.

Bibliography

Balbachevsky, E. and A.J.J. Botelho 2011. Science and Innovation Policies in Brazil: A Framework for the Analysis of Change and Continuity. Paper Presented at the IPSA-ECPR Joint Conference: Whatever Happened to North-South? São Paulo, Brazil, February 2011.

Berman, E. 2014. Not Just Neoliberalism: Economization in US Science and Technology Policy. *Science, Technology & Human Values* 39 (3): 397–431.

Birch, K., and D. Tyfield. 2012. Theorizing the Bioeconomy: Biovalue, Biocapital, Bioeconomics or … What? *Science, Technology & Human Values* 38 (3): 299–327.

Block, F. 2011. Crisis and Renewal: The Outlines of a 21st Century New Deal. *Socio-Economic Review* 9: 31–57.

Borrus, M. 2002. *The Global Pharmaceutical Market: International Trade and Competition in High Technology*, University of California, Haas Business School, Berkeley.

Borup, M., N. Brown, K. Konrad, and H. Van Lente. 2006. The Sociology of Expectations in Science and Technology. *Technology Analysis & Strategic Management* 18 (3–4): 285–298.

Brand, U., C. Görg, J. Hirsch, and M. Wissen. 2008. *Conflicts in Environmental Regulation and the Internationalisation of the State: Contested Terrains.* London: Routledge.

Brown, N. 2003. Hope Against Hype – Accountability in Biopasts, Presents and Futures. *Science Studies* 16 (2): 3–21.

Casper, S. 2007. The German Biotechnology Industry: The Limits of Orchestrated Innovation. In *Creating Silicon Valley in Europe: Public Policy Towards New Technology Industries*, ed. S. Casper, 74–104. Oxford: Oxford University Press.

Chen, Z., H. Wang, Z. Wen, and Y. Wang. 2007. Life Sciences and Biotechnology in China. *Philosophical Transactions of the Royal Society of London* 362 (1482): 947–957.

Chesnais, F. 2004. Globalization Against Development. *International Socialism Journal* 102: 26–44.

Cimoli, M., B. Coriat, and A. Primi. 2008. *Intellectual Property and Industrial Development: A Critical Assessment.* Initiative for Policy Dialogue Working Paper Series. New York: Columbia University.

Clark, G.L. 2000. *Pension Fund Capitalism*. Oxford: Oxford University Press.

Cooke, P. 2001. New Economy Innovation Systems: Biotechnology in Europe and the USA. *Industry and Innovation* 8 (3): 267–289.

Cooper, M. 2008. *Life as Surplus: Biotechnology and Capitalism in the Neoliberal Era*. Seattle: University of Washington Press.

Coriat, B., and O. Weinstein. 2009. *Intellectual Property Right Regimes, Firms and the Commodification of Knowledge*. Comparative Research in Law & Political Economy, Research Paper No. 17/2009. Available at http://digitalcommons.osgoode.yorku.ca/clpe/130. Accessed 20 May 2016.

Coriat, B., F. Orsi, and O. Weinstein. 2003. Does Biotech Reflect a New Science-Based Innovation Regime? *Industry and Innovation* 10 (3): 231–253.

Coriat, P., and G. Schméder. 2006. Post-Fordism in a More Globalized Capitalism. In *The Hardship of Nations: Exploring the Paths of Modern Capitalism*, ed. P. Coriat, P. Petit, and G. Schméder, 311–340. Cheltenham: Edward Elgar.

Daly, H.E., and J.B. Cobb. 1989. *For the Common Good*. Uckfield: Beacon Press.

Drahos, P., and J. Braithwaite. 2002. *Informational Feudalism: Who Owns the Knowledge Economy?* London: Earthscan.

EC (European Commission). 1996. "Biotechnology: State of Play in Europe" Press Release 11th January 1996. Available at http://europa.eu/rapid/press-release_MEMO-96-2_en.htm. Accessed 5 April 2017.

Fortun, M. 2012. Genomics Scandals and Other Volatilities of Promising. In *Lively Capital: Biotechnologies, Ethics and Governance in Global Markets*, ed. K. Rajan, 329–353. Durham, NC: Duke University Press.

Gompers, P. 1994. The Rise and Fall of Venture Capital. *Business and Economic History* 23 (2): 1–26.

Heller, M., and R. Eisenberg. 1998. Can Patents Deter Innovation? The Anticommons in Biomedical Research. *Science* 280 (5364): 698–701.

Hester, R.J. 2016. Biology as Opportunity: Hybrid Rule from a Molecular View. In *Hybrid Rule and State Formation: Public-Private Power in the 21st Century*, ed. S.L. Hurt and R.D. Lipschutz, 175–192. London: Routledge.

Hopkins, M.M., P.A. Martin, P. Nightingale, et al. 2007. The Myth of the Biotech Revolution: An Assessment of Technological, Clinical and Organisational Change. *Research Policy* 36 (4): 566–589.

Hurt, S.L. 2016. What's at Stake in the Privatization Debate? Enclosing the Public Domain Through Hybrid Rule. In *Hybrid Rule and State Formation: Public-Private Power in the 21st Century*, ed. S.L. Hurt and R.D. Lipschultz, 101–122. London: Routledge.

Jasanoff, S. 2005. *Designs on Nature: Science and Democracy in Europe and the United States*. Princeton, NJ: Princeton University Press.

Jessop, B. 2000. The State and Contradictions of the Knowledge-Driven Economy. In *Knowledge, Space, Economy*, ed. J.R. Bryson, P.W. Daniels, N.D. Henry, and J. Pollard, 63–78. London: Routledge.

Johnson, C. 1982. *MITI and the Japanese Miracle: The Growth of Industrial Policy: 1925–1975*. Stanford, CA: Stanford University Press.

Landes, W.M., and R.A. Posner. 2004. *The Political Economy of Intellectual Property Law*. Washington, DC: AEI Press.

Mazzucato, M. 2015. *The Entrepreneurial State: Debunking Public vs. Private Sector Myths*. London: Anthem Press.

Merges, R.P. 2000. One Hundred Years of Solicitude: Intellectual Property Law, 1900–2000. *California Law Review* 88 (6): 2178–2240.

Orsenigo, L. 1989. *The Emergence of Biotechnology: Institutions and Markets in Industrial Innovation*. London: Pinter.

Pagano, U., and A. Rossi. 2009. The Crash of Knowledge Economy. *Cambridge Journal of Economics* 33: 665–683.

Pisano, G.P. 2006. *Science Business: The Promise, the Reality and the Future of Biotech*. Boston, MA: Harvard Business School Press.

Porter, T. 1999. Hegemony and the Private Governance of International Industries. In *Private Authority and International Affairs*, ed. A.C. Culter, V. Haufler, and T. Porter. Albany: State University of New York Press.

Rasmussen, N. 2014. *Gene Jockeys: Life Science and the Rise of Biotech Enterprise*. Baltimore, MD: John Hopkins University Press.

Reinert, E.S. 2009. *Emulation vs. Comparative Advantage: Competing and Complementary Principles in the History of Economic Policy*. Working Papers in Technology Governance and Economic Dynamics No. 25, Tallinn University of Technology.

Safrin, S. 2004. Hyperownership in a Time of Biotechnological Promise: The International Conflict to Control the Building Blocks of Life. *The American Journal of International Law* 98: 641–685.

Senker, J. 1996. National Systems of Innovation, Organizational Learning and Industrial Biotechnology. *Technovation* 16 (5): 219–229.

Shubber, K., and N. Bullock. 2015. US Biotech Bull Run Sets Scorching Pace. *Financial Times*, July 17. Available at http://www.ft.com/cms/s/0/f33ffd50-2adb-11e5-8613-e7aedbb7bdb7.html#axzz3jpSMpX00

Smith, C. 2000. China Rushes to Adopt Genetically Modified Crops. *New York Times*, October 7. Available at www.ag.ohio-state.edu/ipm/trans/10_071.htm. Accessed 25 January 2017.

Smith, T. 2010. Technological Dynamism and the Normative Justification of Global Capitalism. In *Political Economy and Global Capitalism: The 21st Century, Present and Future*, ed. R. Albritton, B. Jessop, and R. Westra, 25–41. London: Anthem Press.

Weiss, L. 2014. *America Inc*. Ithaca, NY: Cornell University Press.

Zeller, C. 2007. From the Gene to the Globe: Extracting Rents Based on Intellectual Property Monopolies. *Review of International Political Economy* 15 (1): 86–115.

CHAPTER 4

Intellectual Property for Pharmaceuticals and Plant Genetic Resources in Historical Perspective

When the General Agreement on Trade and Tariffs (GATT) Uruguay Round was launched in September 1986, its crowded agenda included an item on incorporating negotiations on trade-related aspects of IP, an item that appeared almost as a footnote, with unclear legal foundations and amidst uncertainties that it would survive the end of the Round (Adede 2003). At the end of the Round in 1994, erected as one of the three pillars of the multilateral trade regime stood the footnote item in the form of the TRIPS Agreement, which, apart from supporting the new edifice of the trade regime, single-handedly ushered in the much more restrictive global IP regime of today. Countries are obliged for the first time in history to respect the high IP protection standards demanded by TRIPS, irrespective of the very different socio-economic conditions prevailing in them. As discussed, far from it being enacted so as to stimulate R&D, transfers of technology and economic growth in all its signatories, TRIPS was an effort on the part of the US and the EU to protect the competitive advantage of their most technologically advanced sectors, which, incidentally, had more or less authored it (May 2000; Sell 2003; Cimoli et al. 2008). For the first time in history, frontier states' pursuit of technological rents on behalf of their most promising sectors came fully equipped with (the WTO) arbitration and sanction mechanisms, which could and have been used—as both Brazil and India soon found out[1]—to discipline those playing 'unfairly' in the global marketplace. That the rest of the world was

V. Muzaka, *Food, Health and the Knowledge Economy*, Building a Sustainable Political Economy: SPERI Research & Policy,
DOI 10.1057/978-1-137-59306-1_4

playing 'unfairly' became one of the favourite explanations for the growing trade deficit and perceived loss of competitiveness in the US during the 1970s and 1980s (Destler 1992). As the inability to remedy the trade deficit shifted the political focus to external trade policy, the belligerent stance that came to dominate the latter from the mid-1970s onwards was soon noted by observers of US domestic affairs, followed by those trade partners who got an early taste of the US newly enacted unilateral sanctions for their 'unreasonable' barriers to American trade, and finally by the entire GATT membership called at the US behest to the negotiating table for a new round for which they had little appetite in the early 1980s (Destler 1992; Ostry 2000).

Of the areas that required a fundamental rebalancing from the US viewpoint—namely, agriculture and manufacture goods, services, investment and IP, all areas in which the US had a significant comparative or absolute advantage—IP and services were most emphatically resisted by most developing countries as being unsuitable for inclusion in trade negotiations (Watal 2001; Adede 2003). But the link between trade and IP was by then not only well established in the US but had also become part of its trade law and practice. As the direct result of mounting pressure from high-tech and copyright industries that laid the blame for the growing US trade deficit abroad and promoted themselves as the vibrant industries capable of improving US competitiveness—more so if their IP assets were 'properly' protected—the US Congress tightened a number of provisions authorising the US President and (later) the Trade Representative Office to take retaliatory trade actions against 'offending' countries in the 1984 Trade Act (Destler 1992; Sell 1999). Introduced and strengthened in the 1974 and the 1979 Trade Act, respectively, the infamous Section 301 included, for the first time, in 1984 the failure to protect IP as an 'unfair and actionable' trade practice. Alarmed by the staggering losses US IP-reliant business sectors claimed to suffer due to inadequate protection abroad—in the grotesquely inflated range of US$43–61 billion reported in 1988 (Emmert 1990: 1327)—the US Congress tightened the screws on Section 301 again in the 1988 Omnibus Trade and Competitiveness Act. Brazil, having already been listed alongside other 'offending' developing countries on the Section 301 watchlist from 1985 onwards, became the test case for the revamped Section 301: faced with reluctance to amend its patent law, the US applied 100% tariff increases on several Brazilian goods the same month Brazil argued that there was no place for IP negotiations at the GATT, save those on easing their restricting effects on access to technology (Ryan 1998). A few months later, Brazil, India and other countries no longer voiced any objections to IP being negotiated at the GATT, thus opening the way for TRIPS—the ultimate leveller—to emerge.

4.1 IP Protection for Pharmaceuticals

As had been the case post-WWII, the US state was once again attempting to tilt the international playing field in favour of sectors that were perceived to hold the greatest promise of enhancing its prospects of growth and competitiveness. That it itself had engaged in similar acts during its catch-up phase had little bearing on it fervently disciplining developing countries now engaged in acts of 'piracy' and 'theft' of the most valuable assets of these sectors: IP titles. Neither had the fact that claims of 'piracy' and 'theft', so successful in raising sentiments of unfairness, were invalid, for some of these countries were not signatories to the existing IP treaties—US itself did not sign the Berne Convention (on copyright) until 1989—and the treaties themselves deemed perfectly legal the national variations in IP protection standards (Henderson 1997). A report by the World Intellectual Property Organization (WIPO) prepared for the GATT IP negotiations in 1988 revealed the scale of such variations in patent protection: out of 98 members of the Paris Convention (on patents), 49 excluded pharmaceutical products from protection, 44 excluded plant varieties, 35 excluded food products and 42 excluded biological processes for animal and plant varieties (WIPO 1988). Many developed countries had themselves taken advantage of international IP flexibilities to limit pharmaceutical patents to processes only; Switzerland, with a pharmaceutical sector as competitive as any, introduced patents for pharmaceutical products only in 1977, while in Italy and France, the ban on drug patents was completely lifted only in 1978. Not only could countries exclude entire areas from patentability prior to TRIPS, but patent duration when they did offer it also varied; for instance, pharmaceutical patent duration varied from 5 years in some developing countries to around 15 years in developed ones (WHO 2010). After TRIPS came into force, not only in pharmaceuticals, but in all areas of technology, patents had to be granted without exception and for no less than 20 years.

The more patent duration (re)gained its importance as a strategic tool to protect surplus profits for proprietary pharmaceutical companies, the more they mobilised to extend it. Soon after the patents' 'Dark Ages' came to an end in the US, such mobilisation bore fruit in the form of the 1984 Hatch–Waxman Act—followed ten years later by similar measures in the EU—which extended the patent term for an extra five years due to regulatory delays (Danzon and Keuffel 2005), a euphemism for the time during which public health authorities check the safety of the new drugs or medical devices. A new layer of protection in the form of data exclusivity protection—effectively treating as trade secrets pharmaceutical research data contained in the registration file submitted to public authorities—was also included in the 1984 Hatch–Waxman Act; coming in 2004, the EU's response was late but twice as generous, granting data exclusivity for up to ten years (Dinca 2005). Once these and other innovative extensions of IP protection terms entered the US regulatory sphere, they became the measuring stick against which 'unfair' practices abroad could be identified and assessed, regardless (or perhaps because) of TRIPS having already extended the patent term to 20 years and recognised the importance, but not mandated specific terms, for data exclusivity. To the chagrin of many developing countries, acquiescing to TRIPS—which, in the words of one of the key private sectors' strategists, granted them '95% of what we wanted' (Jacques Gorlin, in Sell 2003: 160)—brought more rather than less Section 301 pressure to bear on them, especially on account of their 'unsatisfactory' IP protection for pharmaceuticals (Muzaka 2011).

That proprietary pharmaceutical companies are especially reliant on IP protection—from their perspective because of an unfortunate combination of a costly R&D process with its easily codifiable nature that makes their products particularly vulnerable to copying—is evident in the extent of their historical involvement with the patent system to ensure it meets their requirements (Macdonald 2002; Angell 2004). Showing remarkable consistency, among the high-tech and copyright sectors to whose mobilisation TRIPS owes its emergence and content, the proprietary pharmaceutical sector played a crucial role, and consequently, in this area of patentability, changes brought about by TRIPS were among the most radical (Ryan 1998; Drahos and Braithwaite 2002; Sell 2003). The timing of the expansion of pharmaceutical IP titles—geographically, temporally and qualitatively—is closely linked to broader developments accompanying the shift towards post-Fordism discussed so far, but also the state of the pharmaceutical sector during the 1970s and 1980s. During this period, compared with the declining fortunes of other industrial sectors in the West and especially of the chemical sector to which it was historically linked, the proprietary pharmaceutical sector emerged as nothing less than a 'sunrise sector': its higher-than-average profit margins ensured it an uninterrupted top position among Fortune 500 companies, whereas its worldwide market more than doubled from US$70 billion in 1975 to US$150 billion (in constant 1980 US$) in 1990 (Ballance et al. 1992; Scherer 1993). Throughout this period, the North American, European and Japanese markets—where almost all proprietary pharmaceutical companies are headquartered—accounted for around 80% of their global sales, with the comparatively unregulated US market alone accounting for about 60% of the sector's worldwide profits (EFPIA 2004).

Like the biotech sector discussed earlier, the proprietary pharmaceutical sector owes its success less to its unusually innovative and entrepreneurial streak than to the many public hands that keep it afloat. On the demand side, new institutional arrangements established post-WWII in the form of public healthcare systems or mixed private–public insurance schemes created a more or less guaranteed market for medicines, granting the sector a level of independence from the vagaries of the business cycle of which companies in other sectors could only dream (Ramirez and Tylecote 1999). Importantly, on the supply side, too, new medicines and processes would not have been forthcoming but for the substantial contribution made by universities and major support from substantial public funds in the US and Europe. Starting with the undisputable success of the German pharmaceutical companies when the modern pharmaceutical sector was coming into being during the late 1800s, followed by the rise of US pharmaceutical sector post-WWII and the re-emergence of the European pharmaceutical sector during the 1950s and 1960s, were all achieved to a considerable extent through close cooperation with academic institutions and the transfer of large sums of public money towards funding research (Achilladelis 1999; Ramirez and Tylecote 1999; Angell 2004).

Ironically, patents' 'Dark Ages' were also the golden age of pharmaceutical innovation, boasting around 93 new molecular entities introduced annually on average during the 1960s; the stronger IP protection for pharmaceuticals became from the 1980s onwards, the lower this number fell, down to 44 in the 1990s and around 25 during the noughties (Borrus 2002; CMR 2010). Regardless of attenuated innovative outcomes, the number of pharmaceutical patents continues to increase. As the proprietary pharmaceutical companies found that their high profit margins could be better maintained through playing the patent game rather than increasing their R&D investment, the preferred response to the dark clouds gathering over the innovation pipeline horizon from the early 1990s onwards came in unprecedented waves of mergers and acquisitions (M&As) activity, often engulfing the most promising dedicated biotech firms that had mushroomed in the meantime. Incidentally, claims of a biotech revolution notwithstanding, the emergence of these firms primarily altered—but did not disrupt—existing models of drug development heuristics (Hopkins et al. 2007), and none of them rose to challenge the position of major, established proprietary pharmaceutical firms. On the contrary, hardly constrained by what little remained of the earlier anti-trust measures, the ten largest companies—all of which were headquartered in frontier economies—increased significantly their share of the global pharmaceutical market (by sales) from 12% in 1987 to 48% in 2000 (Danzon et al. 2003: 7).

Even if not economically significant on their own—they would become so during the economic boom of the noughties—developing countries' pharmaceutical markets were nonetheless a critical frontier to bring under control, all the more critical as pharmaceutical patent protection was unevenly granted or not granted at all. It was not African countries—whose rising populations had plenty of health afflictions but whose low purchasing power made them an uneconomic market—but rather developing countries in Asia and Latin America that were of concern, especially those that had managed to develop productive capabilities of their own in the generic pharmaceutical sector. As many of them could only produce final products out of active pharmaceutical ingredients, the 'pirate' label was applied more forcefully to the few, namely Argentina, Mexico, India, China and South Korea, that could manufacture generic versions of patented pharmaceuticals through reverse engineering (Kaplan and Laing 2005). Brazil was of no immediate concern on this front. Despite the weakening of pharmaceutical patent protection through changes to Brazil's patent law in 1945 and 1971, banning pharmaceutical product patents in the former and banning pharmaceutical patents altogether in the latter (Mazzoleni and Póvoa 2009), foreign direct investment (FDI) flows in the sector increased, a confounding phenomenon that latter-day TRIPS advocates could not explain. But these flows were almost exclusively directed at buying domestic firms; partly as a result of proprietary pharmaceutical companies' strategies to gain a foothold in the large Brazilian and Latin American market, and partly because of Brazil's much more dependent developmental model, the domestic pharmaceutical was effectively taken over by foreign companies. The share of foreign control of the sector increased from 13.5% in 1930 to nearly 80% in 1969 (Ackerman 1971: 21). The rapid increase in FDI flows during the 1970s, more rapid than in other sectors of the economy, sealed its control by foreign pharmaceutical companies by the end of that decade (Gereffi 1983). Despite such level of control—or perhaps because of it—Brazil was singled out as an 'offending' country on the pharmaceutical IP front as soon as the new 1984 US Trade Act included IP offences as actionable under Section 301. As will be discussed in the sixth chapter, this and other events had significant repercussions for the trajectory of IP reform in Brazil from the early 1990s onwards.

At the end of a protracted process that had started soon after independence—and on the basis of two comprehensive reports, the Chand Report (1950) and the Ayyangar Report (1959), both pointing at the link between the patent protection afforded to foreign pharmaceutical companies and drug prices among the highest in the world—emerged in 1970 India's new patent law that banned pharmaceutical product (but not process) patents (Mueller 2008). In marked contrast to the Brazilian case, following the new patent act and a whole raft of governmental policies that included restrictions on foreign ownership, discriminatory use of government procurement and fiscal incentives, many foreign pharmaceutical companies exited the Indian market (Bhattacharjea and Sindhwani 2014). The vibrant domestic sector that grew as a result of these measures succeeded not only in bringing down drug prices significantly in India—although not low enough for its poorest—but also in becoming an internationally competitive sector that earned India the label of 'pharmacy of the world' in some quarters and that of the worst kind of 'pirate' in others. Unsurprisingly, the complaint against India's patent law provisions was initiated barely a year after TRIPS came into force by the US (later joined by the EU) on behalf of their proprietary pharmaceutical companies (Reichman 1998; Matthews 2002). This was the first-ever WTO TRIPS dispute[2]—thus also the first dispute between states over IP to be adjudicated at the international level—and the first to be subjected to the entire arbitration procedures the WTO had been equipped with. As will be seen in the next chapter, once the dust after the historic Panel and the Appellate Body decisions settled, it became clear that India's domestic stance on pharmaceutical patents had not been unequivocally vindicated.

In the wake of these early signals, the earlier international patent regime and all the flexibilities it had afforded appeared progressive to many developing countries, which soon found the TRIPS 'one-size-fits-all' patent regime too restrictive. But, like the latter-day regime, the earlier international patent regime was not created to assist developing countries' progress, and this had been the reason why they had mobilised from the 1960s onwards to change it. For, as formal colonial rule consolidated during the nineteenth century, the international patent regime also expanded to include the colonies not with their interests in view, but rather to secure the colonisers' economic interest against each other in colonial territories (Ugolini 1999; Okediji 2003). Despite being completely alien to the socio-economic realities over which they exerted their power, patent laws were not annulled after decolonisation. On the contrary, mediated by the same international institutions that had facilitated their colonisation earlier on, most newly independent states adhered to inherited IP statutes as a 'privilege' and 'duty' of statehood (Drahos 2002; Okediji 2003). To be sure, some developing countries—India and Brazil among them—did make changes to their inherited IP laws during the 1970s along the lines discussed above. None of these reforms was heretical, however, each conforming to the rules of the international IP regime of the day and in line with (IP) institutional practices found in many frontier economies. As will be discussed in the third section, even the most formidable international mobilisation on the part of developing countries to change international patent rules during the 1960s and 1970s unfolded entirely within the contours of the existing international patent regime and, in retrospect, only succeeded in preparing the way for TRIPS' arrival.

4.2 IP Protection for Plant Genetic Resources

While mobilised in an effort to relax extant international patent rules so as to improve access to technology during the 1960s and 1970s, another front was being opened in the emergence of new forms of IP protection for plant varieties which, due to timing, developing countries could not have inherited but would come under increasing pressure to introduce. Before biotechnology would decisively eliminate various biological barriers to the commodification of seeds[3] and thus intensify the subsumption of nature in the service of value generation, the emergence of hybridisation as a new method of plant (*in*)breeding in the early twentieth century had already opened the way for controlling plant reproductive capacities and turning (certain) seeds—as distinct from the final produce—into commodities (Boyd et al. 2010). This was because hybridisation provides its own built-in mechanism of protection: as the hybrid loses its vigour after one generation, farmers are obliged to purchase new seeds (and much else besides) every season (Raustiala and Victor 2004). Without a scientific understanding of germplasm—the genetic material that contains 'the code' of the inherited traits of an organism—thousands of unnamed farmers had been using plants' natural reproductive proclivities to alter the genetic make-up of crops and bequeath them freely to posterity since prehistoric times. Now a technology appeared that would contribute to farmers' separation from their means of production—the seed being both the means of production and, in the end, the produce (Kloppenburg 2010)—and to the strengthening of the plant breeding and seed sector, especially from the 1930s onwards. That hybridisation, the new method of breeding that emerged in the US initially in corn and later in all crops amenable to this technology, became the technology of choice from the 1930s onwards had less to do with its technological superiority than with the commercial interests of private seed breeders and the legal and financial institutional arrangements put in place by the US state (Kloppenburg 2004). In time, these measures led to the renegotiation of private–public boundaries in germplasm research that, not unlike the biotech sector later on, resulted in the emaciation of previously robust public research institutions—incidentally, where the corn hybrid crosses first developed—not only in the US, but in Europe, too (Kloppenburg 2004). As this private sector grew, so did its demand for protection of its most important asset: scientific knowledge of the selection and testing processes of high-performing hybrids and the successful final hybrids themselves. A new form of IP protection—plant breeders' rights (PBRs)—was developed to enable this new enclosure and extend protection for plant breeders in local and international markets, materialising first in the International Convention for the Protection of New Varieties of Plants (UPOV) signed in 1961 by six European countries, followed by the US adopting its own Plant Variety Protection Act in 1970.

An active and flourishing horticultural sector had in fact mobilised since the late 1800s for a *plant* patent system in the US, joined later in its efforts by seed companies. Despite Congress' concerns to eliminate the 'discrimination' between plant developers and industrial inventors, the 1930 Plant Patent Act only covered asexually propagated species (i.e. vegetative ones such as fruit trees and ornamentals), thus excluding a vast array of plants and, importantly for organised farmer groups, food crops (Petit et al. 2001; Roa-Rodriguez and Dooren 2008). Nonetheless, followed by similar acts in Europe (e.g. Germany in 1933 and Austria in 1938), this act signalled the arrival of plant variety development as a commercial sector for which PBRs would become one of the main means of enclosing and appropriating economically valuable germplasm. European commercial breeders, having been no less interested and organised than their American counterparts, achieved earlier success: their efforts throughout the 1950s had resulted in a set of basic principles for protecting (vegetative *and* reproductive) materials of plant varieties that was eventually adopted in the 1961 UPOV (Dutfield 2011). This success, in turn, revived commercial breeders' efforts to achieve the same protection in the US, which came in the form of the 1970 Plant Variety Protection Act.[4] Importantly, PBRs were made compatible with the notion of germplasm as a 'common heritage of mankind'; because the locus of economic value was the plant variety or the seed, that is, a particular combination of genes that had been manipulated through breeding and later through genetic engineering to create a new variety, it was this that was removed from the commons and privatised via PBRs, whereas the 'raw material', that is, germplasm, remained open and public (Raustiala and Victor 2004; Roa-Rodriguez and Dooren 2008). The uniform varieties that had been so manipulated came to be known as 'elite' cultivars, whereas the heterogeneous landraces/germplasm from which they had been extracted, as 'raw' or 'primitive', occluding the fact that the latter, too, were the outcome of improvement efforts on the part of traditional breeders, that is, peasant and indigenous communities, over millennia (Kloppenburg 2010) .

It is difficult to overstate the importance of UPOV: before it, there were no IP titles over the appropriated and 'worked' germplasm. The PBRs offered by the 1978 revised UPOV, which became the international standard, did not match patent rights—although the 1991 UPOV would come close enough[5]—but they nonetheless became a potent tool in commercial breeders' hands to enter, extend and protect their markets. All UPOV versions (1961, 1972, 1978 and 1991) provide breeders with the exclusive rights to produce commercially, offer to sale and market protected plant varieties, the latter being protected for up to 25 years on account of them being new, distinguishable, uniform and stable.[6] As seed companies succeeded in acquiring IP property titles, they became attractive to other large firms operating in cognate sectors such as pharmaceuticals, chemical and grain trading companies, occasioning a shift from a market characterised by thousands of smaller firms to a highly concentrated one. For instance, as agrochemical companies took advantage of relaxed anti-trust measures to effectively take over the seed industry from the 1970s onwards, they also combined into fewer and fewer companies: what had been 30 separate agrochemical companies in the 1970s became

Monsanto, DuPont, Syngenta, Bayer, Dow and BASF—the Big Six—which by the end of the first decade of the new century controlled 60% of global proprietary seed market and 76% of the global agrochemical market (Howard 2015: 3). Following a series of regulatory and legislative changes initially in the US from the 1980s onwards that opened the way to financial speculation and hedging in agricultural commodity markets, this sector, too, would become highly financialised (Murphy et al. 2012).

These tendencies were no doubt exacerbated by the promise of biotech in the late 1970s and particularly by the 1985 decision of the US Patent Office Board of Appeals in the *Hibberd* case on genetically engineered corn, which, taking its cues from the 1980 *Chakrabarty* case, opened the gates to seed/plant patents. As a result, patents on genetically modified (GM) seeds increased rapidly, as did their control by top seed companies: the top three companies (Monsanto, DuPont and Syngenta) owned no less than 70–80% of GM seed patents 20 years after the *Hibberd* case (Glenna and Cahoy 2009: 124). Such concentration in patent ownership has unsurprisingly gone hand in hand with the rather constricted character of agro-biotech: over 40 years since Cohen and Boyer came up with the rDNA technique and following the commercialisation of the first GM seed in 1996 in the US, a rather narrow range of GM seeds—soybean, corn, cotton and canola claiming the lion's share—incorporate only two agronomically relevant GM traits: herbicide tolerance and Bt (*Bacillus thuringiensis*) insecticidal action. Still, the GM crop acreage grows: since 2011, more than half of global GM crop area is in the developing world, with Brazil leading the table—second only to the US—followed by Argentina, India and China (James 2015).

It is a cruel trick of nature—following its own logic that typically resists obliging human intentions—to have arranged for most of the planet's germplasm to be located in areas that would eventually come under the sociopolitical label of 'the South'. Around 95% of plant genetic resources that provide the bases of the global food crop production originated in regions of 'the South', especially the Latin American and West Central Asiatic regions (Kloppenburg 2004: 181). Long before the emergence of hybridisation and genetic engineering that shifted the locus of economic value onto the plant variety, plant *species* with high commercial value (e.g. rubber, cotton, coffee, tea) had been the loci of prolonged conflicts among great colonial powers striving to gain commercial and military advantages (Petit et al. 2001). The creation of a worldwide network of botanic centres by the European states in the eighteenth and nineteenth centuries was a clear manifestation of the intensification of an 'imperial botanic chess game' that had started with the Columbian Exchange in the fifteenth century (Crosby 1972). During this imperial game, plant germplasm was appropriated from regions now labelled the developing world, its commercial utility and viability ascertained, and subsequently transferred to plantations in European colonial possessions across the world (Crosby 1972; Mooney 1983). Such transfer contributed significantly to the European states' economic fortunes by providing the genetic foundations of their plantation cash crops abroad and, indirectly, through lowering the costs of food necessary for the social reproduction of their swelling working population at home (Mooney 1983; Kloppenburg 2004). Importantly, it also contributed directly to the patterns of socio-economic development and underdevelopment in the developing world—including India and Brazil—that are still in play today.

One visible consequence of such appropriation of germplasm is the competitive advantage of EU, US[7] and Australian agribusiness in the global market that is based wholly on crops that originated from the developing world (wheat, corn, soy etc.). Another, and for our purposes more important, consequence of the imperial game in germplasm manifested itself during the late 1970s, when many developing countries, having already seen genetic resources bountiful in their territories return to them as commodities, were coming under pressure to adopt private PBRs in the UPOV mould (Mooney 1983). The preferred PBRs model was UPOV because the rights it afforded to commercial breeders—all of which were at this time located in frontier economies—offered them significant tools to control the seed market that was expanding rapidly as the vast but uneconomic market of the developing world was being converted, thanks to the Green Revolution, into a highly profitable one. Officially legitimised on account of feeding the world and protecting forests from felling, the Green Revolution was a key moment in the process of the subsumption of nature under Fordism: apart from the geopolitical aim of stemming the 'communist danger' in developing countries, its aim was nothing less than the transformation and incorporation of their pre-capitalist agrarian social formations into the orbit of capital accumulation (Kloppenburg 2004; Patel 2007).

In this process of reconfiguring relations to the means of production and altering the balance of social forces, the state, in its national and international forms, was again central: the state made the Green Revolution and the Green Revolution remade the state and society in specific ways (Patel 2013: 18). Neither its inception nor its rollout to parts of the developing world would have been possible without state intervention, most notable in the form of donor states' foreign development aid policies, initially from the US and later also from international organisations, especially the World Bank. Hybrid corn—having become the backbone of the American agriculture and accounting for half of the world corn production during the first part of the twentieth century (Warman 2003: 181–183)—became the emblem not only of hybridisation but also of the Green Revolution early on, overtaken by wheat later. In world grain markets, American corn and wheat were dominant, a trend exacerbated by US foreign aid programmes that resolved the crisis of US agricultural overproduction by flooding developing countries with dumped grains, displacing domestic production there and making food dependency a chronic phenomenon (Fowler 1994; Warman 2003; Patel 2013). The state played an equally important role in making the Green Revolution in participating developing countries visible in the numerous subsidy programmes created to realise Revolution's promises of economic growth, in legislative measures enabling the extension of the Revolution and of capitalist agriculture domestically, and, importantly, in efforts to repress or stabilise the attendant social discontent that inevitably erupted (Warman 2003; Kloppenburg 2004; Patel 2013). Far from being ingenuous victims of the Green Revolution, the aim of these states was the transformation of their 'backward' peasant agriculture into a more capital-intensive mode of production, often as part of their modernisation, nation-building and development strategies of the post-war/decolonisation period.

At least two consequences of these radical transformations are worth highlighting with respect to plant genetic resources. First, the international institutional network of the Green Revolution served not only as a mechanism to upend traditional farming methods and the social structures in which they were embedded in order to create the necessary conditions for industrial agriculture, but it also served as a mechanism to systematise and deepen the appropriation and transfer of plant genetic resources from these countries to the industrialised ones (Kloppenburg 2004). The accelerated collection of plant genetic resources was carried out during this period largely under the supervision of the Consultative Group on International Agricultural Research (CGIAR)—a private–public, donor-led group with weak oversight by the UN Food and Agriculture Organization (FAO), to which it is linked—the modern successor of the eighteenth- and nineteenth-century botanical centres network (Mooney 1983; Kloppenburg 2004). By the early 1990s, this process had resulted in around 85% of the genetic material thus collected being deposited in the gene banks located in the North and the CGIAR's own research centres (Fowler 1994). Large seed/agrochemical companies also continued to expand their own, private seed banks.

Second, and related, the extension and intensification of the Green Revolution practices simultaneously necessitated a constant supply of 'primitive' germplasm, whose genetic variability was necessary for the supply of commercially viable hybrids, and directly contributed to the loss of such genetic diversity. Concerns about the Revolution and its considerable ecological effects were made more systematically from the 1960s onwards, while those related to genetic erosion reached the FAO in the early 1970s (Brand et al. 2008). Genetic erosion occurred not only because once a hybrid replaced a variety in the field, the latter, and its germplasm, was lost, but also because other varieties and hybrids in the market tended to mimic the commercially successful one. In other words, higher-yield varieties eliminated the sources from which they emerged (Wilkes 1984). Adding to this contradictory mix the environmental concerns that swept across many domestic and international policymaking fora during the 1970s, the position many developing countries found themselves can be summarised thus: as it became clear that the expansion of industrial agriculture simultaneously eroded genetic resource diversity and exposed the importance of control over them, many developing countries came under pressure to both guarantee access to plant genetic resources in their territories under the principle of 'common heritage of mankind' and to offer private PBRs for 'elite' cultivars penetrating their domestic markets. Growing unease led to efforts to organise the flow of plant genetic resources during the 1980s, a process that, as will be discussed below, effectively ended the 'common heritage' system without resolving the social, economic and environmental problems accompanying the accelerated appropriation of plant genetic resources.

4.3 International Contests over Patents and Plant Genetic Resources Prior to TRIPS and the CBD

Nearly a year before Mexico announced that it was no longer able to service its debt in 1982—widely accepted as the beginning of the debt crisis—and roughly at the same time as the Reagan administration embarked on significantly expanding funding for the US life science sector, speaking on behalf of a group of developing countries, Mexico demanded that the FAO prepare a draft international agreement for controlling the flow of plant genetic resources, amidst threats to stop germplasm transfer from the South (Petit et al. 2001; Brand et al. 2008). What became known as the 'seed wars' had thus begun. Although the debt crisis that erupted in 1982 would send many of them in a downward social and economic trajectory for a decade or more, at least two events in this period gave developing countries the reason to believe that their hope of creating a new international economic order in their favour could still be realised. One of them materialised out of 'seed wars' at the FAO in the form of the 1983 International Undertaking on Plant Genetic Resources.[8] Acquiring only a voluntary status, this Undertaking was not the binding convention developing countries had wanted, but carried significant symbolic power: the principle of 'common heritage' had been confirmed and it applied, as developing countries had insisted, not only to 'primitive' germplasm but also to 'elite' commercial cultivars/lines (Mooney 1983; Kloppenburg 2004). The second appeared at the WIPO, where a separate 'war' over international patent rules and access to technology had been raging between developed and developing countries for some time and which in 1982 seemed to be settling in favour of the latter, save the thorny issue of compulsory licensing[9] (Watal 2001; Drahos 2002). In hindsight, these fleeting victories may have blurred developing countries' representatives' vision of the radical transformations occurring in frontier economies that were ushering in the emergent post-Fordist regime of growth. Ten years later, the 'common heritage' principle was as good as dead for 'primitive' and 'elite' germplasm alike, while TRIPs set patent standards diametrically opposed to those that developing countries had been demanding, including a new obligation to introduce a *sui generis* system of protection for plant varieties—often in practice meaning UPOV—and much else besides.

Representing a number of developing countries concerned about the restrictive effects of patents on access to technology, Brazil officially inaugurated the patent reform movement by submitting a proposal on patent protection and transfer of technology to underdeveloped countries at the UN in 1961.[10] Underpinned by a latent demand for restitution for past colonialism, this movement sought to reform the rules of the international IP regime in order to facilitate the growth of developing countries' economies. The patent reform movement, alongside similar efforts spearheaded by India to reform the international copyright system with a view to responding to developing countries' concerns about the spread of knowledge, literacy and mass education, was eventually subsumed under the broader movement of the G77 for a fairer New International Economic Order (NIEO) (Murphy 1983; Yu 2009). As it would turn out, this demand for fairness was worlds apart from that the US state was gearing up to achieve during the same period. But even though developing countries' proposals to reform the copyright and patent rules so as to facilitate their development strategies appeared threatening to frontier states, they did not represent a heretical way of managing knowledge production and/or circulation. On the contrary, they were primarily concerned with reforming the existing rules of the game, especially the technology and knowledge transfer rules, without seeking to radically upset the IP regime of the day (Muzaka 2013).

Even in this they did not succeed, notwithstanding the fact that their chances looked bright against the backdrop of economic recession and the crisis of Fordism in the frontier economies during the 1970s. Contests over copyright reform precipitated a crisis in international copyright in the late 1960s, followed by skirmishes at the UNESCO (United Nations Educational, Scientific and Cultural Organization), which ended with the withdrawal of the US and the UK from it in 1984 (Drahos and Braithwaite 2002). Contests over patents and technology transfer splintered into negotiations over revisions of patent rules at the WIPO and negotiations about an International Code of Conduct on the Transfer of Technology at the UNCTAD (United Nations Conference on Trade and Development), both of which recognised patents rights and were hardly radical. The US did not withdraw from the WIPO, but following the collapse of negotiations in 1984, it decisively moved IP negotiations to the GATT, whereas the UNCTAD Code was buried in 1985. The global IP regime that emerged in the wake of these failed efforts—and the new but not fairer economic order of which it became part—could not have been further away from what developing countries' representatives had in mind.

As efforts to create a new international economic order based on fairness and solidarity were practically out of steam by the mid-1980s, principles that the G77 had defended vigorously during the NIEO movement were also being abandoned. Two important principles emerged on the issue of natural resources out of the NIEO movement. One came in the form of a 1962 UN General Assembly Resolution[11] that confirmed the sovereignty of nations over their natural wealth and resources in the wake of increased global demand for raw materials, whose exploration rights remained largely in the hands of foreign companies, a vestige of colonial times. The second came in the form of calls for the application of the 'common heritage of humankind' principle—with roots in Roman law precepts of common-space resources as well as in the public trust doctrine—to ocean and seabed resources in 1967, as a response to increased conflicts over access to and benefits from their exploitation.[12] As for their impact on genetic resources, these two principles underpinned conflicting and shifting alliances during the 1970s and 1980s. Frontier states insisted that 'primitive' genetic resources were the common heritage of humankind but resisted the application of this principle to oceans and seabed resources, whereas most developing countries insisted that both genetic resources and seabed resources were common heritage of humankind, only to extend the principle of permanent sovereignty to genetic resources within their territories by the end of the 1980s.

The most visible but by no means the only forum where this shift took place was the FAO itself. The 1983 Undertaking was but a momentary truce: although voluntary in nature, it was seen to threaten the principle of private property rights over 'elite' cultivars by the US and a number of other frontier economies with a competitive advantage in the seed and agrochemical sectors. The 'seed wars' resumed and, in hindsight, only abated when three resolutions were agreed over the interpretation of the FAO Undertaking in 1989 and 1991,[13] when TRIPS negotiations at the GATT were in full swing. The paltry attention these 'agreed interpretations' drew at the time was the exact opposite of their significance: all states, including developing ones, simultaneously recognised sovereign rights over genetic resources within their territories and the private rights of commercial plant breeders, throwing in as a counterbalance of sorts the recognition of the historical contribution made by traditional farmers in the form of non-binding Farmers' Rights (Kloppenburg 2004; Safrin 2004). An international fund was set up to support the latter, but the pot remained more or less empty (Brand et al. 2008: 111). In the meantime, an agreement between Costa Rica and the (US) pharmaceutical company Merck granting the latter exclusive access to and IP rights over genetic resources collected in Costa Rica in return for a one-off payment of US$ 1 million signalled not only the end of the 'common heritage' principle, but also the demise of the unity developing countries—now gearing up to compete with each other for similar commercial agreements—had showed on this front up until then.

Further pressures were emerging simultaneously from a different and unlikely front: contests over international environmental issues (Raustiala and Victor 2004). As environmental concerns grew and the one-at-a-time approach to addressing habitat and species loss met its limits, a broader approach towards protecting entire ecosystems gained ground and finally materialised in the Convention on Biodiversity (CBD) in 1992. Saving the tropical rainforests—an especially popular political issue in developed countries during the 1980s—offered them an opportunity to couple this special concern with access to plant genetic resources and biodiversity more generally (McAfee 1999; Roa-Rodrigues and Dooren 2008). Perceiving this as yet another intrusion into their sovereignty, developing countries reacted by expanding the principle of national sovereignty from natural resources to include forest resources, plant genetic resources and biodiversity overall (Raustiala and Victor 2004). Unlike the FAO Undertaking, by virtue of being a legally binding convention, the CBD succeeded in establishing a new property domain over plant genetic resources limited and ruled by state sovereignty. Simultaneously, with the interests of seed, pharmaceutical and agrochemical companies in mind, the US and other frontier economies made sure that the CBD recognised the exclusive private rights of those in the business of manipulating genetic material, although clearly not strongly enough for the US, which signed but did not ratify the CBD partly on this account (Rosendal 1994). Farmers' Rights, even in non-binding form, found no place in the CBD, although some references to the rights of 'indigenous and local communities' were made (Kloppenburg 2004). Apart from this significant omission of one of the key issues over which the 'seed wars' were fought at the FAO, the CBD left unresolved the status of the considerable collections kept in the international gene banks set up before it came into force, the economic importance of which was another decisive impetus for the start of the 'seed wars' (Brand et al. 2008). The CBD simply recognised the special status of plant genetic resources for agriculture and established for them—and for genetic resources more generally—a property domain that simultaneously recognised the sovereign rights of the state and the private property rights of business (Petit et al. 2001). In other words, raising more expectations than solutions on the issue of plant genetic resources for agriculture, the CBD put the ball back in FAO's court.

As conflicts over IP multiplied from the mid-1990s onwards in all these fora, it was not the conflicts emerging between the CBD and the FAO rules on plant genetic resources for agriculture, but rather those between the CBD and TRIPS that stole the show. This was perhaps unsurprising, given the status of each in the international regime hierarchy, as was the fact that TRIPS would usually take precedence, being the only agreement that is both legally binding and equipped with an arbitration and sanction mechanism. This said, the CBD–TRIPS conflict is not a substantial one, because both are based on an understanding that genetic resources are economically valuable resources. This element is clearly visible in TRIPS, of course, but also in the CBD, which simultaneously enshrines the sovereign right of states over genetic resources and links inextricably their conservation to utilisation. Not only does the CBD not restrict access to what is to be protected, but it facilitates it; reflecting the new-found faith in market mechanisms, conservation of biodiversity according to the CBD is only possible through their use, thus codifying the primacy of the economic utility and value of genetic resources over their preservation (Kloppenburg 2004; Brand and Görg 2013). It was not negotiation fatigue or miscalculation but rather this economic utility—enhanced by the promise of biotechnology already hyped up in frontier economies as the means of turning DNA into gold—that contributed to developing countries rich in biodiversity radically shifting their position regarding genetic resources. In the context of wider transformations taking place in the global economy during the 1970s and 1980s that ushered in a more intensive regime of valorisation and a concomitant shift in the role of the state, developing countries, too, adopted versions of the competitive orientation that had emerged in frontier economies. The manner in which this orientation found expression with regard to the institutional forms of IP in pharmaceuticals and agro-biotech in Brazil and India is discussed in more detail in the following chapters.

Notes

1. Of the 24 TRIPS-related disputes brought before the WTO by 2001, 22 were brought by the US and the EU separately or together; reference in this case is made to the case against India brought by the US in 1997 on account of the former not having complied with TRIPS provisional arrangements for pharmaceutical and agricultural chemical patents, especially the so-called mailbox arrangement for filing patent applications until India fully complied with TRIPS in 2005. The US brought a case against Brazil in 2000 on account of certain 'local working' provisions in its patent law, which was settled bilaterally. For more, see Abbott (2009).

2. Technically, the cases by the US against Pakistan and Portugal, both filed in April 1996, came earlier than the Indian one (July 1996), but the former two were eventually withdrawn (see https://www.wto.org/english/tratop_e/dispu_e/dispu_status_e.htm).
3. Seed is used here as a generic term, referring to the part of the plant that is used for propagation. This includes both sexual seeds ('true' seeds, the result of ovule fertilisation) and vegetative seeds (products of cuttings, grafting, tissue culture or special plant parts used for propagation such as tubers, roots or bulbs).
4. US commercial breeders achieved more than their European counterparts because, unlike the European version that demanded new cultivars to be demonstrably superior, the 1970 Plant Variety Protection Act in the US placed no burden on them to improve quality, resulting in the clear growth of commercial breeding sector from 1970 onwards, without an increment of yield gain over the historic trend established by public breeders up until then (Kloppenburg 2004: 142).
5. Currently, 74 nations are UPOV signatories: most (54) have ascended to the 1991 version since the 1978 version closed to new signatories in 1999. The 1991 version of the UPOV, meant to put the UPOV system on nearly equal footing as the patent system, grants significantly greater rights to plant breeders. The two main exceptions to PBRs are the breeders' exemption and the farmers' privilege. The former does not allow breeders to prevent other breeders from creating new varieties based on their protected varieties. The latter enables farmers to use the seeds (and other propagating materials) of protected plant varieties for non-commercial purposes without the breeders' prior authorisation. However, the 1991 UPOV prohibits farmers growing protected varieties to sell the seeds they harvest from the crop and, increasingly in many UPOV member countries, from saving and exchanging seeds on a non-commercial basis. Likewise, the 1991 UPOV restricted the breeders' exemption; those using a protected variety to create a new variety have to make major changes to the genotype, lest their new variety is considered an 'essentially derived' variety, falling into the ownership of the first breeder.
6. The novelty criterion, added in 1991, is not the same as the novelty criterion for patentability but merely means that the variety has not been previously commercialised. The other criteria for a plant variety to be protected are its distinguishability, meaning that a variety of rice, for example, is different from any other variety; its uniformity, meaning that all the plants in question should display the same characteristics; and its stability, meaning that the variety should display the same characteristics in each successive generation. In UPOV 1978, the term of protection was 15 years (but at least 18 years for varieties of trees and wines), whereas in UPOV 1991, this was extended to 20 years and 25 years, respectively.

7. US golden age of germplasm collection started later, in the early 1900s.
8. FAO 1983 International Undertaking on Plant Genetic Resources: follow the link to the Undertaking at http://www.fao.org/nr/cgrfa/cgrfa-about/cgrfa-history/en/.
9. A compulsory license is issued by a public authority, usually the same office that grants the patent, forcing the patent-holder to license their protected good or technology to others at compensation, for example, for the reason of public interest. It has been a feature of the international patent system since the Paris Convention was signed, but the grounds on which compulsory licensing can be issued have become much more narrow in practice.
10. UN General Assembly (1961). *The Role of Patents in the Transfer of Technology to Under-Developed Countries.*, Proposal submitted by Brazil on 8 November 1961. UN Doc. A/C.2/L.565 and A/C.2/L.565 /Rev.1.
11. Resolution 1803, Permanent Sovereignty over Natural Resources.
12. These discussions would eventually lead to the 1982 UN Law of the Sea Convention, which came into force only in 1994.
13. See FAO Resolutions 4/89, 5/89 and 3/91 at http://www.fao.org/Ag/cgrfa/iu.htm.

Bibliography

Abbott, F. 2009. TRIPS Dispute Settlement Decisions. Presented at International Centre for Trade and Sustainable Development (ICTSD) Seminar on 26–27 October 2009, Switzerland. Available at http://www.ictsd.org/downloads/2009/10/abbott_trips_dsu.pdf. Accessed 5 May 2016.

Achilladelis, B. 1999. Innovation in the Pharmaceutical Industry. In *Pharmaceutical Innovation – Revolutionizing Human Health*, ed. R. Landau, B. Achilladelis, and A. Scriabine. Philadelphia: Chemical Heritage Press.

Ackerman, F. 1971. Industry and Imperialism in Brazil. *Review of Radical Political Economics* 3: 1–39.

Adede, A.O. 2003. Origins and History of the TRIPs Negotiations. In *Trading in Knowledge: Development Perspectives on TRIPs, Trade and Sustainability*, ed. C. Bellmann, G. Dutfield, and R. Melendez-Ortiz, 23–35. London and New York: Earthscan.

Angell, M. 2004. *The Truth About the Drug Companies: How They Deceive Us and What to Do About It.* New York: Random House.

Ballance, R., J. Pogany, et al. 1992. *The Worlds Pharmaceutical Industries: An International Perspective on Innovation, Competition and Policy.* Aldershot: Edward Elgar.

Bhattacharjea, A., and F. Sindhwani 2014. *Competition Issues in the Indian Pharmaceuticals Sector.* Working Paper, Centre for Development Economics, Delhi School of Economics, New Delhi, India. Available at http://www.cuts-ccier.org/compeg/pdf/report-pharmaceutical_sector_study.pdf. Accessed 20 May 2016.

Borrus, M. 2002. *The Global Pharmaceutical Market: International Trade and Competition in High Technology*, University of California, Haas Business School, Berkeley.

Boyd, W., S. Prudham, and A. Schurman. 2010. Industrial Dynamics and the Problem of Nature. *Society & Natural Resources: An International Journal* 14 (7): 555–570.

Brand, U., and C. Görg. 2013. Regimes in Global Environmental Governance and the Internationalization of the State: The Case of Biodiversity Politics. *International Journal of Social Science Studies* 1 (1): 110–122.

Brand, U., C. Görg, J. Hirsch, and M. Wissen. 2008. *Conflicts in Environmental Regulation and the Internationalisation of the State: Contested Terrains.* London: Routledge.

Cimoli, M., B. Coriat, and A. Primi. 2008. *Intellectual Property and Industrial Development: A Critical Assessment.* Initiative for Policy Dialogue Working Paper Series. New York: Columbia University.

CMR (Centre for Medicines Research). 2010. *The CMR International Pharmaceutical R&D Factbook.* Chicago: Thomson Reuters.

Crosby, A. 1972. *The Columbian Exchange: Biological and Cultural Consequences of 1492.* Westport, CT: Greenwood Press.

Danzon, P., and E. Keuffel. 2005. *Regulation of the Pharmaceutical Industry.* Paper Presented at the National Bureau of Economic Research Conference on Regulation, 9–10 September 2005, Cambridge, MA.

Danzon, P., A. Epstein, et al. 2003. *Mergers and Acquisitions in the Pharmaceutical and Biotech Industries.* Organizational Economics of Health Care Conference 2003. Available at http://ssrn.com/abstract=468301. Accessed 30 May 2016.

Destler, I.M. 1992. *American Trade Politics.* New York: Institute for International Economics and the Twentieth Century Fund.

Dinca, R. 2005. The "Bermuda Triangle" of Pharmaceutical Law: Is Data Protection a Lost Ship? *The Journal of World Intellectual Property* 8 (4): 517–563.

Drahos, P. 2002. Developing Countries and International Intellectual Property Standard-Setting. *The Journal of World Intellectual Property* 5: 765–789.

Drahos, P., and J. Braithwaite. 2002. *Informational Feudalism: Who Owns the Knowledge Economy?* London: Earthscan.

Dutfield, G. 2011. The Role of the International Union for the Protection of New Varieties of Plants (UPOV). Global Economic Issue Publications, Intellectual Property Issue Paper Number 9, Quakers United Nations Office, Geneva. Available at http://quno.org/sites/default/files/resources/UPOV%2Bstudy%2Bby%2BQUNO_English.pdf. Accessed 6 June 2016.

EFPIA. 2004. *Annual Report Brussels, European Federation of Pharmaceutical Industries and Associations.* European Federation of Pharmaceutical Industries and Associations, Brussels, Belgium (in file with author).

Emmert, F. 1990. Intellectual Property in the Uruguay Round – Negotiating Strategies of the Western Industrialized Countries. *Michigan Journal of International Law* 11: 1317–1399.

Fowler, C. 1994. *Unnatural Selection: Technology, Politics and Plant Evolution.* Verdon, Switzerland: Gordon & Breach.

Gereffi, G. 1983. *The Pharmaceutical Industry and Dependency in the Third World.* Princeton, NJ: Princeton University Press.

Glenna, L.L., and D.R. Cahoy. 2009. Agribusiness Concentration, Intellectual Property and the Prospects for Rural Economic Benefits from the Emerging Biofuel Economy. *Southern Rural Sociology* 24 (2): 111–129.

Henderson, E. 1997. TRIPs and the Third World: The Example of Pharmaceutical Patents in India. *European Intellectual Property Review* 11: 651–663.

Hopkins, M.M., P.A. Martin, P. Nightingale, et al. 2007. The Myth of the Biotech Revolution: An Assessment of Technological, Clinical and Organisational Change. *Research Policy* 36 (4): 566–589.

Howard, P.H. 2015. Intellectual Property and Consolidation in the Seed Industry. *Crop Science* 55: 1–7.

James, C. 2015. 20th Anniversary (1996 to 2015) of the Global Commercialization of Biotech Crops and Biotech Crop Highlights in 2015. ISAAA Brief No. 51. Available at http://isaaa.org/resources/publications/briefs/51/executivesummary/default.asp. Accessed 6 June 2016.

Kaplan, W., and R. Laing. 2005. *Local Production of Pharmaceuticals: Industrial Policy and Access to Medicines*. Available at http://siteresources.worldbank.org/HEALTHNUTRITIONANDPOPULATION/Resources/281627-1095698140167/KaplanLocalProductionFinal.pdf. Accessed 10 November 2015.

Kloppenburg, J. 2004. *First the Seed. The Political Economy of Plant Biotechnology, 1492–2000.* 2nd ed. Madison: Wisconsin University Press.

———. 2010. 'Impending Dispossession, Enabling Repossession: Biological Open Source and the Recovery of Seed Sovereignty. *Journal of Agrarian Change* 10 (3): 367–388.

Macdonald, S. 2002. Exploring the Hidden Costs of Patents. In *Global Intellectual Property Rights: Knowledge Access and Development*, ed. P. Drahos and R. Mayne. Basingstoke: Palgrave Macmillan.

Matthews, D. 2002. *Globalising Intellectual Property Rights: The TRIPs Agreement.* London: Routledge.

May, C. 2000. *A Global Political Economy of Intellectual Property Rights: The New Enclosures?* London: Routledge.

Mazzoleni R., and L.M.C. Póvoa. 2009. Accumulation of Technological Capabilities and Economic Development. Available at http://ideas.repec.org/f/pma824.html. Accessed 4 October 2016.

McAfee, K. 1999. Selling Nature to Save It? Biodiversity and Green Developmentalism. *Environment and Planning D: Society and Space* 17: 133–154.

Mooney, P.R. 1983. *The Law of the Seed: Another Development and Plant Genetic Resources*. Uppsala: Dag Hammarskjöld Foundation.

Mueller, J. 2008. *Biotechnology Patenting in India: Will Bio-Generics Lead a Sunrise Industry to Bio-Innovation?* Legal Studies Research Paper Series, University of Pittsburgh School of Law, Working Paper No. 2008-2.

Murphy, C.N. 1983. What the Third World Wants: An Interpretation of the Development and Meaning of the New International Economic Order Ideology. *International Studies Quarterly* 27 (1): 55–76.

Murphy, S., D. Burch, and J. Clapp. 2012. *Cereal Secrets: The world's Largest Grain Traders and Global Agriculture*. Oxfam Research Report. Available at https://www.oxfam.org/sites/www.oxfam.org/files/rr-cereal-secrets-grain-traders-agriculture-30082012-en.pdf. Accessed 6 June 2016.

Muzaka, V. 2011. *The Politics of Intellectual Property Rights and Access to Medicines*. Basingstoke: Palgrave Macmillan.

———. 2013. Contradictions, Frames and Reproductions: The Emergence of the WIPO Development Agenda. *Review of International Political Economy* 20 (1): 215–239.

Okediji, R. 2003. The International Relations of Intellectual Property: Narratives of Developing Country Participation in the Global Intellectual Property System. *Singapore Journal of International and Comparative Law* 7: 315–385.

Ostry, S. 2000. Looking Back to Look Forward: The Multilateral Trading System After Fifty Years. In *From GATT to the WTO: The Multilateral Trading System in the New Millennium*, ed. WTO Secretariat, 97–112. The Hague: Kluwer Law International and the WTO.

Patel, R. 2007. *Stuffed and Starved: Markets, Power and the Hidden Battle for the World Food System*. London: Portobello Books.

———. 2013. The Long Green Revolution. *The Journal of Peasant Studies* 40 (1): 1–63.

Petit, M., C. Fowler, et al. 2001. Why Governments Can't Make Policy: The Case of Plant Genetic Resources in the International Arena. Paper prepared for the Global Forum on Agricultural Research. http://agris.fao.org/agris-search/search.do?recordID=XF2016008984. Accessed 6 June 2016.

Ramirez, P., and A. Tylecote 1999. *Technological Change in the Pharmaceutical Industry: A Literature Review from the Point of View of Corporate Governance*. COPI Report Prepared for EC DG XII. Sheffield University, UK.

Raustiala, K., and D. Victor. 2004. The Regime Complex for Plant Genetic Resources. *International Organization* 58 (2): 277–309.

Reichman, J.H. 1998. 'Securing Compliance with the TRIPs Agreement after US vs. India. *Journal of International Economic Law* 1 (4): 585–601.

Roa-Rodriguez, C., and T. Dooren. 2008. Shifting Common Spaces of Plant Genetic Resources in the International Regulation of Property. *The Journal of World Intellectual Property* 11 (3): 176–202.

Rosendal, K. 1994. Implications of the US 'No' in Rio. In *Biodiplomacy: Genetic Resources and International Relations*, ed. V. Sánchez and J. Calestrous, 87–103. Nairobi: African Centre for Technology Studies.

Ryan, M.P. 1998. *Knowledge Diplomacy: Global Competition and the Politics of Intellectual Property*. Washington, DC: Brooking Institution Press.

Safrin, S. 2004. Hyperownership in a Time of Biotechnological Promise: The International Conflict to Control the Building Blocks of Life. *The American Journal of International Law* 98: 641–685.

Scherer, F.M. 1993. Pricing, Profits and Technological Progress in the Pharmaceutical Industry. *The Journal of Economic Perspectives* 7 (3): 97–115.

Sell, S. 1999. Multinational Corporations as Agents of Change: The Globalization of Intellectual Property Rights. In *Private Authority and International Affairs*, ed. A.C. Cutler, V. Haufler, and T. Porter, 169–198. Albany: State University of New York Press.

Sell, S. 2003. *Private Power, Public Law: The Globalization of Intellectual Property Rights*. Cambridge: Cambridge University Press.

Ugolini, M.A. 1999. Grey Market Goods Under the Agreement on Trade-Related Aspects of Intellectual Property Rights. *Transnational Law* 12: 451–468.

Warman, A. 2003. *Corn and Capitalism: How a Botanical Bastard Grew to Global Dominance*. Trans. by N.L. Westrate. Chapel Hill: University of North Carolina Press.

Watal, J. 2001. *Intellectual Property Rights in the WTO and Developing Countries*. The Hague: Kluwer Law International.

WHO. 2010. Intellectual Property Protection: Impact on Public Health 2010. http://www.who.int/medicines/areas/policy/AccesstoMedicinesIPP.pdf. Accessed 14 May 2016.

Wilkes, G. 1984. Germplasm Conservation Toward the Year 2000. In *Plant Genetic Resources: A Conservation Imperative*, ed. C. Yeatman, D. Kafton, and G. Wilkes, 131–163. Boulder, CO: Westview Press.

WIPO. 1988. Existence, Scope and Form of Generally Internationally Accepted and Applied Standards/Norms for the Protection of Intellectual Property, issued as GATT document MTN.GNG/NG11/W/24/Rev1. Geneva WIPO.

Yu, P.K. 2009. A Tale of Two Development Agendas. *Ohio Northern University Law Review* 35: 465–573.

CHAPTER 5

Intellectual Property for Pharmaceuticals and Agro-Biotech in India

One of the issues on which governments of various political orientations in post-1991 India have unanimously agreed is that the development of the biotech sector is in India's national interest. Shortly after laying down his vision of transforming India into a knowledge society in 2000, Prime Minister (PM) Atal Bihari Vajpayee stated at the annual Indian Science Congress in Delhi that India's global vision included 'shaping biotechnology into a premier precision tool of the future for creation of wealth' (quoted in Newell 2003: 23). Mirroring the high expectations attached to it in frontier economies, biotech is widely seen among India's policymakers as a 'technology of hope', a powerful enabling technology that would not only revolutionise agriculture and health in India, but also help establish it as a knowledge superpower in the world (GoI 2007). Although imaginaries of biotech as a technology of hope would come visibly to the fore during the 1990s, the Indian state had paid close attention to developments in biotechnology and intellectual protection trends in the US and other frontier economies throughout the 1970s and 1980s. Having already been impressed by the value of genetic manipulation in producing the wheat and rice hybrids that came to symbolise its embrace of the Green Revolution during the 1960s, it responded to technological advances in biotech discussed in the previous chapters by including this emergent sector as a new priority area in its 6th Five Year Plan (1980–1985), a commitment that has only strengthened since then (Rajan 1994; Kumar et al. 2004).

V. Muzaka, *Food, Health and the Knowledge Economy*, Building a Sustainable Political Economy: SPERI Research & Policy,
DOI 10.1057/978-1-137-59306-1_5

Of the important measures that followed this move, the Long-Term Plan in Biotechnology of 1983 made clear the need to strengthen indigenous capacities in biotech in general and its contribution to the agriculture and health sectors in particular, towards which end the National Biotechnology Board was set up in 1982, later incorporated as the Department of Biotechnology (DBT) under the Ministry of Science and Technology (Rajan 1994; Ramanna 2006a). Its early orientation towards the commercialisation of biotechnology, alongside its strong plant breeding programme and wealth of biodiversity/germplasm, underpinned India's reservations about the 'common heritage of mankind' principle during the mid-1980s and its decisive move towards the principle of national ownership from then onwards (Rajan 1994).

Perhaps one of the clearest signs of the Indian state's eagerness to fully exploit the commercialisation of biotechnology was the considerable planning dedicated towards the 1990 Biotech Consortium India Ltd. under the remit of the DBT as a public–private undertaking financed largely by public financial institutions (e.g. the Unit Trust of India and the Industrial Development Bank of India). What is particularly interesting about the Consortium is the manner in which it represented an institutional innovation that was directly inspired by the institutional arrangements that had enabled the emergence and commercialisation of biotech in the US: its aims of financing and fostering close relationships between private biotech firms, research institutes and universities so as to develop and bring biotech products to the market were in line with those the US state had achieved through the 1980 Bayh–Dole Act and its institutional support for the early entry of private capital in the biotech sector. While it cannot be held as an exemplary success, the Consortium's emergence is also notable on account of it simultaneously expressing continuity with the significant role the Indian state has traditionally granted to S&T, and a break in the manner in which their contribution to India's socio-economic fortunes were to be realised. Like the Keynesian states' investment in 'big science' during Fordism in frontier economies, but responding to the specific socio-economic and political conditions in India, the post-colonial state gave S&T a privileged role in (re)making India a great power, often inspired by a narrative that depicted India as having been a knowledge power of sorts in the past (e.g. bestowing the modern number system, astronomy, Ayurveda medicine etc.), stifled in no small measure by colonialism (Rajan 2006; Alamgir 2009). Standing apart from other anti-colonial movements at the time, that which developed in India had a strong intellectual current which granted S&T a key role in nation-building and economic development. This commitment to S&T and its link to nationalism and development, perhaps most clearly espoused by Nehru's conviction that 'it is science alone that could solve [these] problems of hunger and poverty' (quoted in Krishna 1997: 237) and that the future belonged to those 'who made friends with science' (quoted in Visvanathan and Parmar 2002: 2714), contributed to science becoming an essential activity of the post-colonial Indian state.[1] It made considerable investments in 'big science', scientific R&D and high-tech sectors such as atomic energy and aerospace—although, importantly, not in basic education—investments that were unusually high and lopsided for a country with relatively high poverty and illiteracy rates, but which were fully in line with the state's ambition of promoting India's industrial development and its rise to greater global prominence (Krishna 1997; Mahtaney 2007; Alamgir 2009).

5.1 On the Economic Imaginary of (Re)making India a Knowledge Economy

The shift that occurred in the orientation of the Indian state during the 1980s and early 1990s towards internal and external economic liberalisation threw into sharp relief both continuities and changes in the role S&T was to play in improving India's fortunes. The clearest break with the past was not the retreat of the state from supporting and funding S&T: although a relatively low figure of less than 1% of GDP, total R&D spending in the country is largely financed by public R&D funding, which still accounts for around 70% of the total (Bound 2007: 14). Nor did the state's ultimate goal of making India a great power change. Instead, the most important shift is to be found in the manner in which S&T was meant from this point onwards to contribute to India becoming a competitive knowledge power, an imaginary in line with that adopted by frontier economies in post-Fordism. The adoption of the competitiveness orientation and of the knowledge economy as a state project by a core executive–technocratic elite is observable in efforts by the latter to create a popular mobilising imaginary that would legitimise and naturalise this shift by appealing to two contradictory ideas constitutive of the identity of the Indian state that had earlier on justified the inward-looking economic strategy of post-colonial India. These were the global vision of India rightfully recognised as a *world power*, the so-called Indian exceptionalism[2] and *swadeshi*, an idea of various interpretations, all of which insist upon the value of the local over the remote (Kilhani 1997; Jenkins 2003; Alamgir 2009). Against a background of conflicting nationalisms always present in post-colonial societies but thrust more violently into the Indian public sphere from the late 1980s onwards (observable, for our purposes, in the appropriation of Gandhian ideas of moral economy/*swadeshi* by the Hindu Right in the Bharatiya Janata Party's [BJP] programme 'Our Swadeshi Approach: Making India a Global Economic Power'), there was remarkable unity among members of the core executive elite about the necessity of the change in direction. As India's Foreign Minister put it in 1991, 'Defence forces are, of course, essential for protecting sovereignty ... but they are blunt and clumsy instruments ... the market place is the battlefield of the future' (Solanki, quoted in Alamgir 2009: 81), a remarkable statement in a country with a strong intellectual and popular tradition that is suspicious of the market (Jenkins 2003). This was followed, perhaps more tellingly, by Manmohan Singh, one of the key architects of the reforms, proclaiming that 'India's economic destiny is safe only when India knows how to stand on its own feet, to compete against everyone in the world on an equal footing. That is what we are trying to do' (cited in *The Financial Times*, 6 October 1995). Yashwant Sinha, the BJP Finance Minister, confirmed that 'swadeshi actually means competition, going out in the world and winning ... I understand swadeshi basically as a concept which will make India great. And India can be great only when we become an economic superpower ... we can be great by being able to compete' (1998, quoted in Nayar 2000: 807) .

It is during the shift in direction engineered in the late 1980s and 1990s that, as the imaginary of India becoming a competitive market player took hold, the modernisation of the Indian IP system came to be seen as a necessary step in turning India's accumulated intellectual and scientific heritage into growth and wealth (Drahos 2010; Muzaka 2015). An IP system had existed in India since 1856 and, like many of the structures left behind by the coloniser, was incorporated within the new state without any radical overhaul. But up to this point in time, much like the Keynesian states in the frontier economies, the Indian state had sought to harness knowledge's contribution to economic growth not through the commodification and commercialisation of knowledge via IP, but primarily through considerable public investment in 'big science'. Despite criticism of what was later termed the 'Hindu rate of growth' averaging around 3.5% per annum from the 1950s until the 1980s (Herring 1999), this considerable investment in S&T not only propelled industrial growth during the early decades, but was fundamental to the emergence of sectors that would push India into a higher-growth trajectory from the 1980s onwards, including the pharmaceutical and biotech sectors, to which we will return shortly (Rudolph and Rudolph 1987; Kohli 2006). Of the complex sources and nature of the liberalisation reforms that enabled the shift in direction during the 1980s and 1990s, two features demand special attention. The first, as indicated so far, is the new competitiveness orientation and the embrace of the knowledge economy imaginary; being an orientation, it entailed neither the abandonment of lower-value-added sectors, nor an out-and-out liberalisation which materialised relatively slowly and cautiously in practice (Kohli 2006, 2012). But the break with the past orientation was significant and, importantly, was based in no small measure in the hope that the scientific, technological and industrial capacities developed until then would provide the basis on which to compete in high-tech sectors. As the Minister of State noted in 1996, the time had come for India 'to use the industrial base build up mainly for the domestic market over the preceding 30 years to move out into world markets' (Chaturvedi, quoted in Drahos 2010: 213). Bound into this vision of becoming a competitive knowledge economy was the reform of the Indian IP regime so as to enable *Indian* firms to benefit from knowledge monopolies and extract global rents, an IP regime that, as the Secretary of the Department of Scientific and Industrial Research, R.A. Mashelkar, stated in 1996, would turn India's 'intellectual prowess into knowledge and wealth' (quoted in Drahos 2010: 220) .

The second important feature of the shift in orientation of the Indian state in the 1980s and 1990s is that it was the direct result neither of pressures coming from strong business interest groups, nor from powerful international organisations. It is true that India was then (and remains) one of the main clients of the World Bank, and that twice in the space of ten years (1981 and 1991) it had to apply for an IMF (International Monetary Fund) loan, but in both cases, the Indian state only selectively implemented the policy/liberalisation reforms, and importantly, in both cases, many of those undertaken were already planned by the government at the time (Nayar 2000; Randeria 2003; Kohli 2012). In other words, the IMF was pushing against an open door (e.g. in the form of the Rao–Singh plan of 1991) and neither IMF/World Bank, nor the US—aware of changes taking place internally—exerted excessive pressure on the Indian government (Bhagwati 1993; Nayar 2000; Rodrik and Subramanian 2004; Mukherji 2012). Likewise, having seen their fortunes flourish during a period when the Indian state was at the driving seat of the economy as manager of industrialisation and economic growth, the majority of large Indian businesses were not pressing for reforms. As internal and rather modest liberalisation reforms were being launched in the 1980s, it appeared that both business and the state were united in their appreciation of the necessity for more rather than less state, a state that would increase its spending, tighten its grip on labour and support business more actively (Chibber 2012). Beyond this, there existed formidable support for the status quo by a delicately balanced but battle-tested coalition of dominant proprietary classes, including heavyweights such as rich farmers and industrialists (Jenkins 2003). A new business group that had emerged and developed mainly as subcontractors and suppliers to the big powerhouses found its growth prospects limited and therefore was interested in liberalisation. But with fissures within and between business groups intensifying in response to external and continuing internal liberalisation measures, there was limited concerted action by business in support of what became known as the 'big bang' reforms of the 1990s (Alamgir 2009; Chibber 2012).

The orientation towards a more open and competitive economy was instead engineered by the Indian state, or, more accurately, by a powerful fragment within it. A core executive–technocratic elite, including L.K. Jha, Abid Hussain, Shankar Acharya, Montek Singh Ahluwalia and the long-serving Manmohan Singh, played a fundamental role in formulating and pushing the reform process forward from behind the scenes (Mukherji 2012; Kohli 2006; Nayar 2000). In other words, the vision of India as a competitive knowledge economy was a state project, an elite coup formulated within the Indian state and carried out by a small group of reformers initially on a surprisingly thin support base. Well aware of this fact and of the strong resistance against this new orientation, this core group played a formidable role in manipulating and fragmenting political resistance through a number of underhand means, such as semi-legal union busting, playing the pro-reform industry group CII[3] against the more traditional FICCI,[4] various corrupt and illegal practices and, importantly, pushing key reforms through administrative fiat, introducing them as 'statements' rather than as resolutions that would have necessitated parliamentary approval (Jenkins 1999, 2003; Randeria 2003; Chibber 2012; Mukherji 2012). Some of these strategies, as will be discussed, were also used in the transformation of the domestic IP regime that came to be framed as a crucial tool in gaining an advantageous position in the competitive game for economic growth.

As a state project, the economic imaginary of India becoming a competitive global player and a knowledge economy would not only provide some direction and coherence in various fields of state activity, but also be central to legitimising this new orientation and mobilising popular support towards achieving it. Even if the latter cannot be confidently said to have been secured, it is a remarkable fact that governments of different political persuasions have shown unwavering commitment to this orientation (Alamgir 2009; Randeria 2007; Mukherji 2012). Concerns about the ability of the existing technological base to help India succeed in the increasingly competitive world economy already caused a partial shift away from the principle of technological self-reliance in the early 1980s; by the time a new S&T policy was announced in 2003, it became clear that although still nominally important, technological self-reliance had been replaced by the orientation towards global competitiveness as the key goal (Joseph and Abrol 2009). In the same year, and following an exercise in technology forecasting undertaken by the Department of Science and Technology, the government also announced its 'India 2020—Vision for the New Millennium', in which the vision of India becoming one of the top economic powers by 2020 was firmly grounded on advances in technology and innovation (Krishna 2013).

More broadly, the new Indian S&T policy of 2003 also revealed the extent to which the exigencies of competing in the knowledge economy familiar to policymakers in frontier economies had been appropriated by Indian ones: 'The transformation of new ideas into commercial successes is of vital importance to the nation's ability to achieve high economic growth and global competitiveness. Accordingly, special emphasis will be given not only to R&D and the technological factors of innovation, but *also to the other equally important social, institutional and market factors* ... Innovation will be supported in all its aspects ... There is a need to change the ways in which society and economy performs, if innovation has to fructify' (GoI 2003, emphasis added). Alongside attempts such as the Protection and Utilisation of Public Funded Intellectual Property Bill (2008) and the National Innovation Act (2008)—moulded in the fashion of the US Bayh–Dole Act and America Competes Act, respectively—other signs indicative of the orientation on the part of the Indian state towards perpetual innovation and structural competitiveness include the ambitious aims of 'transform[ing] India into an innovative economy' (GoI 2012) and positioning it among the top five global knowledge powers by 2020 (GoI 2003, 2013).[5]

Conceptualising this orientation as a state project does not mean that the Indian state is a Schumpeterian Workfare State. The latter is only an ideal type whose features are differently and unevenly expressed in frontier economies and not immediately applicable as a model for the Indian state. At the same time, as discussed in the second chapter, transformations following the crisis of Fordism have changed the global terrain in which the Indian state must act. And, as noted here, the shift in orientation towards competitiveness had less to do with direct pressures from international institutions or other external actors than with the decision taken within a small but powerful fragment of the Indian state about how India should respond to the changed and changing circumstances. That this core group was more or less successful in occasioning India's change in direction cannot be convincingly explained by arguing that the state in India is more fragmented than frontier states. Certainly, at times a hero or ally, at others a villain or foe, the state in India often appears as a polymorphous creature of many forms (Rudolph and Rudolph 1987). But all modern states are polymorphous (Hobson 1997) and no state has ever been a coherent system whose parts are all dedicated to the same end. The success of a small group in achieving so much can be better understood in light of the nature of a late-industrialiser, post-colonial state such as India's and especially of the manner in which it has historically dealt with the tension all modern states face: that between the national and the global. This relation is everywhere mutually constitutive, but also more fluid and harder to disentangle in post-colonial states (Randeria 2007). For our purposes, the tension at the heart of a post-colonial state like India's can be put thus: while on the one hand, vis-à-vis the peoples that it came to subsume (i.e. 'the nation'), the state is the symbol of independence from colonialism and self-rule (*swaraj*) legitimised by the promise of development which it makes repeatedly, on the other, it finds itself not only unfavourably positioned in global structures, but also dependent on a perception of frontier economies as holding the key to the only real and credible route to development (Vielle 1988; Bergeron 2004).

Since the early days following its formation, a constitutive part of the identity of the independent Indian state was that of engineering and achieving economic development for the nation as a whole via planned *industrialisation*. In other words, the model of development pursued was based precisely on the trodden paths of frontier economies, the only route that would enable India to 'catch up [...] with the Industrial Revolution that occurred long ago in Western countries' (Nehru, quoted in Chatterjee

1993: 202). It sought neither to break up the institutional structures inherited from the colonial past, nor to radically transform the existing social structure. On the contrary, it sought to avoid or neutralise social conflict in the name of its 'rational' pursuit of national development, in the process facilitating both the accumulation of capital which was inevitably at the heart of a developmental model based singularly on the modern industrial sector and legitimising the particularistic interest of such capital invoking the universal value of development and welfare for the nation as a whole (Chatterjee 1993). Negotiating the contradictory space between the global and the nation, the state, whose *raison d'être* was development through industrialisation, thus intervened to enable and facilitate capital expansion and accumulation on behalf of private capital and also on its own behalf as a substantial owner of capital (ibid.). Constituting the main pillar of India's catch-up strategy post independence, inward-looking industrialisation strategies imitated to some extent the Fordist growth model, but, importantly, without the welfare and social protection systems accompanying it in frontier economies. Despite claims of pursuing development for the nation as a whole, the activist–interventionist state acted first as the facilitator and legitimator of capital accumulation, a role that was not only not altered by the liberalisation reforms, but, on the contrary, was strengthen further and extended to include knowledge, the 'new capital' (Muzaka 2015).

With frontier economies perceiving their further growth and competitiveness to depend upon investing in and protecting the (IP) assets of their knowledge-based sectors from the 1980s onwards, it was not an improbable move on the part of the Indian executive elite to also define India's national interests along similar lines. What remains fixed in the significant change in India's economic orientation in the 1990s is this elite group's dependency on perceptions of frontier economies providing the only credible route to India's further development. This said, the path undertaken is a local refraction of economic imaginaries and policies adopted in frontier economies, marked by an ambivalent attitude of attraction and resistance to them. India's embrace of the logic of competitiveness, innovation-led growth, becoming a global knowledge power and so on, unmistakably Western instruments and imaginaries, is coupled with a pronounced sensibility related to neo/colonialism and structural inequalities that contrive to hinder their realisation (Rajan 2006; Randeria 2007). This ambivalence is clear in the case of the global IP regime, seen as an instrument of neocolonialism triggering strong nationalist sentiments, while at the same time adopted as a tool to promote a culture of innovation as the path to realise India's global ambition. Importantly, this ambivalence is suffused by conflicting nationalisms, so characteristic of post-colonial societies, that in India prey on different ideas drawn from a continuum marked by secular–modernist nationalism on the one end and the indigenist, often Hindu, nationalism on the other (Cederlöf and Sivaramakrishnan 2006). As will become clearer in the following two sections, different versions of nationalisms were called upon by the actors involved in the political battles over reforming IP rules for pharmaceuticals and plant biological resources during the 1990s and 2000s.

Towering over them stood the Indian state, seemingly the only actor with the authority to define India's national interest. If the state in the past, acting as the facilitator and legitimator of capital accumulation, contributed to the success of Indian capital being seen as more or less synonymous with national interest (Chatterjee 1993), its closer alliance with Indian business after the reforms could only be expected to strengthen this dynamic. To make a pertinent example in passing, based on the colonial principle of 'eminent domain' that gives the Indian state ultimate control over land and resources—now including genetic resources—within its territory, millions of people have been displaced to open way for large developmental projects (e.g. dams), mining or agriculture without the right to choose as long as the acquisition was said to serve a national or public purpose (Ramanathan 1996). When various social movements started challenging the 'eminent domain' principle, alongside 'public purpose', as a priority of superior order simply confirmed by a statement on the part of the state, the response of the Supreme Court of India was to consistently uphold the sovereign right of the state over its natural resources and, more recently, to interpret the setting up of *private* industry to constitute 'public purpose' (Ramanathan 1996, 2011; Randeria 2003, 2007). This interpretation, in turn, has helped facilitate and legitimise the conversion of common resources into sources of private profit. This conversion lies at the core of IP that the Indian state, as will be discussed in more detail in the following sections, has also embraced on behalf of private capital, as well as its own.

5.2 Reforming IP Rules for Pharmaceuticals

One of the main reasons given for the trajectory of the Indian pharmaceutical sector from one dominated by foreign companies and drug prices among the highest in the world in the early post-independence period to a 'sunrise' sector predominantly controlled by domestic companies and boasting drug prices among the lowest in the world is the ban on pharmaceutical and chemical product patents in the 1970 Patent Act, alongside limited process patent terms (five years) and compulsory licensing provisions aimed at facilitating national technological and industrial development. The 1970 Patent Act was both pragmatic and nationalistic, but it was not distinctive on account of representing a radically different way of governing knowledge production and access to its benefits. On the contrary, it was modelled largely on the 1949 British Patent Act and, as noted in the previous chapter, was wholly within the permissible norms of the international IP system of the day. Internationally, too, the Indian state was not engaged in efforts to radically alter the patent system; its leading role in the international patent reform movement during the 1960s and 1970s was based on the reification of knowledge developed in the West as the only valid one for achieving developmental goals, evidenced in demands that its flow be facilitated within (and through changes to) the existing patent system, demands which not only strengthened indirectly the legitimacy of the international patent regime, but also foreclosed the consideration of alternatives and weakened the basis of resistance to TRIPS from the late 1980s onwards.

What was distinctive about the 1970 Patent Act was the way in which local pharmaceutical firms, generously supported by a number of government policies, for example, import restrictions, tax and price controls and investments in R&D skills, went on to rapidly increase their share of the domestic market from 10% in 1970 to 60% in 1993 and establish themselves as one of the most dynamic and successful sectors in India and globally (Redwood 1994; Mueller 2007; Kapczynski 2009). Initially serving the domestic market, following liberalisation measures from the 1980s onwards, the sector turned its eyes to foreign markets, where prices were higher: before India implemented TRIPS rules for pharmaceuticals in 2005, exports had increased to nearly 40% of total production, up from under 5% in the late 1980s, around 60% of which was accounted for by the US, UK, Germany and France (Greene 2007: 19–20). Before the contribution of the sector would shift to improving India's trade balance and fuelling its hopes of becoming a knowledge economy, the sector's development and success was important to the Indian state not only on account of its commitment to industrialisation, but also in dealing with the issue of drug affordability for the vast majority of the population who could not afford them. In effect, the success of the 1970 Patent Act and other government policies not only helped create a successful industrial sector, but also helped obfuscate the absence of any serious effort on the part of the state to build a robust public healthcare sector.[6] The success of the Indian pharmaceutical sector had created formidably strong constituencies in India against changes to the patent law: the converging interests of a successful generics sector with those of an active group of health non-governmental organisations (NGOs) on the issue of pharmaceutical patents would produce a strong coalition that was simultaneously pro-health and pro-*Indian* business when proposals to bring India's patent rules in line with TRIPS emerged in the 1990s.

Even before the risk of strict and binding international patent rules crystallised in the shape of TRIPS, Indian pharmaceutical sector representatives—for example, the Indian Drug Manufacturers Association (IDMA)—alongside other business groups such as FICCI, vehemently opposed a move by the Indian state to join the (comparatively much laxer) Paris Convention on patents in 1986, arguing it was against India's national interest, a position reiterated when the proposal was floated again in 1988, only one year prior to India conceding to the negotiation of substantive IP issues at the GATT which it had hitherto opposed (Ramanna 2002, 2005). As noted in the previous chapter, although the pressure on India to change its position regarding IP negotiations at the GATT was real and substantial, the thinking within the core executive elite regarding India's economic direction was already shifting by then. This is observable, for instance, in the change in position in the GATT in 1989 having been carried out without the explicit approval of the wider Cabinet, which appears to have been kept in the dark (Raghavan 2003). When the magnitude of the changes in the IP system to which the Indian state was on the verge of committing became clearer in the 'Dunkel Draft' in 1991, domestic criticism of it amounting to 'economic slavery' was strong (Keayla 2005; Ramanna 2005), but except for asking for some additional footnotes, India accepted the draft in 1993.

Such acceptance was initially justified by the government's concerns that opposition would lead to restrictions on India's exports, but key sections of the ruling Congress Party already favoured changes to the patent law by then as part of the wider changes India had to undertake (Ramanna 2002: 2066; Alamgir 2009). After having been elected on a *swadeshi* electoral platform, the same core members of the BJP who had opposed the new IP direction, among them the new PM Atal Bihari Vajpayee, became strong supporters when in government and proceeded to discipline recalcitrant BJP members, whereas central Congress figures, including key reformers and PMs Narasimha Rao and Manmohan Singh, became even keener in opposition to see the new patent regime instituted (Dasgupta 1999; Ramanna 2002). This process was carried out largely through three amendments in 1999, 2002 and 2005 amidst fierce resistance and political contests. Two main strategies were devised to counter such resistance. The first was—and continues to be—that of drumming up support for a 'robust' patent regime in largely apathetic but key scientific and business constituencies, especially through setting up numerous IP awareness programmes and IP facilitating cells across departments, universities and public research institutions, often with the generous help of international organisations such as the WIPO (Muzaka 2015). As a result of these efforts, by the late 1990s, all the major public research institutions—the Council for Scientific and Industrial Research (CSIR), the Indian Council of Agricultural Research and the Indian Council for Medical Research—had in-house units dealing with and promoting IP.

By the mid-1990s, as GDP growth rates picked up and the 1991 crisis appeared a distant memory, arguments about the necessity of accepting TRIPS for fear of retaliation on India's exports were sidelined in favour of arguments that a 'robust' patent system was crucial to India's successful rise in the global knowledge economy. A locus of change along these lines, as well as one of its main instigators, is to be found in the CSIR—one of the world's largest networks of industrially oriented public research labs and, with no less than 40 national research laboratories under its control, the largest in India—under the leadership of R.A. Mashelkar, who, in addition to heading the CSIR from 1995 to 2006, also held a number of key governmental positions related to science, technology and IP. In Mashelkar's decisive hands, CSIR's focus shifted towards engineering India's scientific and technological priorities for a more global market-driven agenda, as well as becoming a significant market player itself. In converting the CSIR into a body that is in tune with India's new knowledge economy project—one symbolism of which is the shift in the CSIR mantra from 'publish or perish' to 'patent and prosper'—Mashelkar and his group appear to have succeeded in collapsing the division between science and technology and redefining the purpose of academic research to serve commercial interests along the same lines as in frontier economies (Rajan 2005, 2007). The CSIR itself claims one of the largest shares of Indian patents granted both in India and abroad, especially in the US (Dutz 2007: 64; Bound 2007: 35).

The second strategy against resistance to changes to the patent regime was a combination of manipulative means and administrative fiat. Because fulfilling the conditions of India joining the WTO required the ratification of TRIPS on 1 January 1995, an ordinance[7] was issued in the last days of December 1994 after the Indian parliament had adjourned with a view to circumventing what were bound to be lengthy and adversarial debates. The many techniques used by the core executive elite (including PM Narasimha Rao) to get the ordinance approved by the parliament in 1995 were devious and wholly undemocratic (Dasgupta 1999); although they failed, they were indicative not only of the strong resistance to the new (IP) course that India was embarking upon, but also of the determination of a small executive–technocratic team to see it through. Apart from resistance from the generics sector, out of the broader movement that emerged when the decisive shift in the government's economic direction towards liberalisation became clear arose a number of NGOs that would become actively involved in the issue of pharmaceutical patent reform. Key players included health NGOs that were part of the Jana Swasthya Abhiyan movement, the Affordable Treatment and Action Campaign and the Lawyers' Collective HIV/AIDS Unit. But perhaps the most significant was the National Working Group on Patent Laws established in 1988 by a small number of prominent lawyers and ex-governmental officials, who, appreciating the finality of the government's shift with regard to IP, used their considerable political and technical acumen to ensure that the new IP regime would safeguard the interest of the generic pharmaceutical sector and the continuity of affordable medicines in the market (Gupta 2010; Matthews 2011).

It is impossible to say what trajectory the patent reform process would have taken had it not been for the US initiating a dispute case against India at the WTO in July 1996—later joined by the EU—on behalf of their proprietary pharmaceutical sectors' complaints about insufficient protection for pharmaceuticals and chemicals in India, especially the mailbox and exclusive marketing rights (EMRs) requirements of TRIPS (Articles 70.8 and 70.9).[8] These transitory rules—the quid pro quo of India and other developing countries having succeeded in buying some time until the full application of pharmaceutical patent rules in 2005—provided proprietary transnational companies with more or less the same level of protection as pharmaceutical patents would have done. The Indian Patent Office was in fact accepting patent applications, but this was done under an *administrative* mailbox mechanism that both the WTO Panel and the Appellate

Body agreed did not amount to an adequate fulfilment of the relevant TRIPS obligations (Reichman 1998), giving an early warning signal against domestic experimentation in implementing them. The decision, in turn, gave the government in Delhi impetus and justification to seek again to amend the patent act it had failed to achieve in 1995. Although generics sector representatives (IDMA) and civil society actors were generally opposed to this renewed effort on account of patent-like effects of EMRs and the mailbox backlog increasing the number of patents post 2005, it is notable that by the time the amendment bill reached the Lok Sabha in the late 1990s, the position of various groups and their relative power had changed. The health NGOs that had strongly opposed the new IP direction early on had become divided and industry bodies' position had also changed (Ramanna 2002). FICCI not only performed a *volte-face* when it supported India's application to accede in 1998 to the same Paris Convention that it had opposed in the late 1980s, but had already established an institute for IP that, among other things, promoted the CSIR-inspired slogan 'patent or perish' to its members (Ramanna 2002: 2067). CII, an early supporter of patent reform, insisted that the strengthening of the patent system in India was paramount to attracting FDI and up-to-date technology, and to improving India's economic fortunes.

Changes had also occurred within the pharmaceutical sector, the most resistant sector to patent reform. Although the TRIPS spectre caused the majority of the bigger Indian generic companies to seek to penetrate developed (especially the US) markets rather than embrace the 'patent race', a small minority also sought to invest in R&D, again, largely targeting developed countries' markets (Rangnekar 2006; Kapczynski 2009). This small group, which included Dr. Reddy Labs and Ranbaxy, joined forces with foreign pharmaceutical companies, arguing that patent reform was fundamental not only to continued drug innovation, but also to the shift from a generic to a modern research-based pharmaceutical sector in India. Some of the credit for these changes belongs to consecutive governments that, as was argued above, had actively promoted India as an important hub in the global knowledge economy and IP as one of the means of making it happen. Indeed, the chairman of Dr. Reddy Labs acknowledged that R.A. Mashelkar had not only inspired scientists 'to create wealth by harnessing IP, but also … all of us in the industry' (Reddy 1999, quoted in Ramanna 2002: 2067).

Other state measures in support of patent reform included capitalising on the wide emotional response that followed the successful challenge of a US patent on turmeric in 1997—discussed in the next section—to show that changes to India's patent laws were not meant to benefit foreign multinationals, but rather to provide protection for *Indian* knowledge, modern and traditional (Kavani 1999). Bound up with the overarching vision of India becoming a prominent global player was the ever-present and now increasing shadow of China's rise[9]; with China signalling that it was reforming its patent system in order to play the patent-based innovation game, the case for patent reform among the Indian core executive elite strengthened (Drahos 2010). By the time India applied to become party to the Paris Convention and the Patent Cooperation Treaty (PCT) in 1998—this time without the fierce domestic opposition it faced ten years ago—China's Intellectual Property Office had already been granted the 'international searching authority' status, which India achieved only in 2007. Despite this lag, India is now among a small league of countries enjoying this status, one that confirms its commitment to the innovation and patent race, as well as to its aim of benefitting from it.

Succeeding on this front demands a difficult balancing act: creating the conditions for and actively promoting an innovation and patenting culture among generic companies, research institutes, universities, governmental departments and public at large on the one hand, and dealing with significant—if fragmented—resistance on the part of sections of civil society and the generics sector on the other. All this has to be achieved while protecting the interest of the generics sector on account of its real and expected contribution both to India's economic performance and to meeting its numerous health needs. One of the ways these tensions materialised in practice was in the Indian state playing a key role internationally in defending the right of sovereign states to give precedence to public health needs over protecting pharmaceutical patents and to design IP laws more in keeping with their diverse national needs, whereas domestically pushing for an IP system that would enhance India's IP assets and help propel the country into becoming a knowledge society/economy. Internationally, for instance, the Indian state actively supported the 2001 WTO Doha Declaration on TRIPS Agreement and Public Health confirming the TRIPS flexibilities to deal with public health concerns (e.g. compulsory licensing); it also played a key role in the following TRIPS amendment process aimed at enabling countries to export drugs produced under compulsory licensing, important in terms of safeguarding the export opportunities of its generics sector even though India had never issued compulsory licenses for drugs before (Muzaka 2011). But despite the importance given to India's leadership in these global IP contests, the government's draft amendment bills during 2002, 2003 and 2004 drafted to make India's patent law fully compliant with TRIPS by 2005 made no mention of the Doha Declaration, placed significant procedural burdens on the issue of compulsory licenses, reduced the scope of pre-grant opposition and were too vague on the issue of patent 'evergreening' (Iyer 2002; Keayla 2005). This was perhaps unsurprising, given that the department in charge of the patent amendment process—the Department of Industrial Policy and Promotion (Ministry of Commerce and Industry), which enjoyed direct contact with the PM office—was particularly keen on improving India's IP assets, a view for which it came into constant conflict not only with NGOs and pharmaceutical sector representatives, but also with other ministries, especially the Health and Environmental ones (Keayla 2005; Mueller 2007).

In the maelstrom of contests accompanying each proposed bill up until the Amendment Act of April 2005—each passed amidst accusations that the process was rushed without proper consultation (Keayla 2005)—the most successful venue that generics sector representatives found was through the Working Group on Patents in collaboration with the Left parties, without whose support the draft bill could not pass through Lok Sabha (Keayla 2005; Basheer 2005). One of the most important issues to the generics sector was that of limiting patents to new chemical entities only, repeatedly rejected by the government as inconsistent with TRIPS; the infamous Section 3(d) that eventually dealt with this issue was introduced only a few days before the bill passed in April 2005. Taking advantage of the fact that TRIPS lists patentability criteria (e.g. novelty) but does not define them, 3(d) prohibits patents on known substances unless of significantly enhanced efficacy, thus providing a legal means to deter patent 'evergreening' and the rentier practices of foreign pharmaceutical companies. For their part, the latter vehemently opposed the section and went on to challenge it through the judicial system, incidentally revamped to include the newly established Intellectual Property Appellate Board created in 2009 in the mould of the US CAFC discussed in the third chapter. The most important case was that of Novartis, whose patent application for its cancer drug Glivec, one of the few to enjoy executive marketing rights in India before 2005, was rejected following a pre-grant opposition in 2006. Novartis then challenged the validity of Section 3(d) as unconstitutional and incompatible with TRIPS in a number of courts; in its decision eagerly anticipated in India and worldwide, the Indian Supreme

Court ruled in April 2013 that Novartis' patent claims were invalid, causing some to praise India for its leadership in IP–health issues and proprietary companies to lament India's hostility towards patents.

Reflecting the difficulty of propelling existing sectors into the knowledge economy game whose constitutive rules are already tilted in favour of those operating at the technological frontier, the split within the Indian pharmaceutical sector among the vast majority of 'imitators' and a minority of *prospective* innovators led to concerns within the government and industry circles that Section 3(d) would place significant hurdles to patenting by *Indian* innovators themselves, including those working on innovations based on Ayurvedic medicines. These concerns were no doubt exacerbated by the impression that India was lagging behind in commercialising its traditional medicines sector compared with China; indeed, the Chinese government had made significant efforts to standardise, modernise and internationalise its traditional Chinese medicine sector, making it a pillar of its innovation-driven economic growth model (Waldmeier 2015). In any case, as 'enhanced efficacy' in the provision remained undefined in practice, the Indian Patent Office has granted a not-insignificant number of pharmaceutical patents in breach of Section 3(d) (Kapczynski 2009; Drahos 2010). While Section 3(d) remains intact until policymakers see it in India's 'national interest' to do so, other measures have already been taken to address its potentially dampening effects on Indian innovators. For instance, although the Indian state took the defensive step against 'biopiracy' by making traditional knowledge non-patentable and digitising large swathes of it as 'prior art' that foreign patent offices can consult before granting patents (Krishnaswamy 2011), it has actively supported the creation of a modern–traditional Ayurvedic medicine sector aimed primarily at Western markets. These 'new' traditional drugs are often patented in India (but not abroad), in line with the fact that the recognition and protection of traditional knowledge by the Indian state is conditional upon it being compatible with its marketing and commodification (Islam 2010; Pordié and Gaudillière 2014).

For those Indian pharmaceutical companies not inclined to delve in Ayurveda or unable to invest in R&D, a new valorisation front was opened through the Indian state's efforts to promote the Indian population as a 'guinea pig' and India as an ideal location for (foreign) pharmaceutical drug trials. These measures included the 2005 patent reform that limited the possibility of patented drugs marketed as generics and the amended 2005 Drugs and Cosmetics Act, which effectively made it permissible for all stages of clinical trials to be undertaken in India (Prasad 2009). In seeking to facilitate collaboration between foreign and Indian pharmaceutical companies, 'advantage India'—a wide gene pool, access to many patients, meager incentives and illiteracy—appeared more like washing dirty linen in public, barely occulting the state's visible hands as a market agent seeking to turn the (otherwise unproductive parts of) Indian population into experimental subjects from which value could be generated (Rajan 2005; Prasad 2009). With an eye on such value, and despite the fact that TRIPS does not demand data exclusivity but the US and the EU offer it to pharmaceutical companies, a proposal emerged within the government for India to do the same in 2007, subsequently shelved on account of considerable opposition by NGOs and the generics sector (IDMA 2009).

The strengthening of pharmaceutical patent protection in India since 2005 deepened the cracks the spectre of TRIPS had caused in the sector. R&D investment in the sector has grown—encouraged not only by changes in patent law, but also other public measures such as the 2004 Pharmaceutical R&D Support Programme, the 2008 Fund for Accelerating Start-Ups in Technology and others, all bearing strong resemblance to similar programmes in frontier economies—but research tends to be concentrated within a few companies, of which Dr. Reddy and Ranbaxy are the most important. (Disappointing nationalists of all kinds, both companies have recently been bought by a Japanese company.) Given the difficulties involved in developing new drugs that even established proprietary transnational companies encounter, most Indian firms investing in R&D have focused these investments in their areas of strength, for example, expanding their production facilities or their generic markets abroad (Greene 2007; Wilson and Rao 2012). At the same time, a contract research industry has grown, driven by measures discussed above and the strategies of proprietary transnational pharmaceutical companies to contract out clinical trials and custom manufacturing services. Hence, despite its technological prowess and rapid economic growth more recently, it appears that India's embrace of the knowledge economy and concomitant efforts to promote innovation in the pharmaceutical sector have so far taken the form of contract work for Western companies, to whom patent rights and benefits largely accrue (Rajan 2006), laying bare once again the persistent structural inequalities that the vision of India becoming a competitive knowledge economy is meant to overcome.

5.3 Reforming IP Rules for Plant Genetic Resources

Demonstrating the importance the Indian government attaches to biotechnology, the very first patent under the new 2005 Patent Act was granted to a biological drug by Roche. Symbolising the contested nature of the new IP orientation in India, the Roche patent came immediately under attack from domestic (civil society and business) groups using post-grant patent opposition procedures and was subsequently revoked. Although pre-/post-grant opposition provisions in the patent law—incidentally, relatively weak in earlier drafts but strengthened following objections from the generics sector and civil society groups—make an Indian patent more susceptible to public interest challenges than elsewhere, their balancing effect depends crucially on the involvement of consistently vigilant and well-resourced civil society groups and a sympathetic juridical system, none of which can be taken for granted. While the courts have so far established a small but important pro-health body of precedence (Shaver 2011; Park and Jayadev 2011), this stance is not guaranteed. Indeed, in its *Dimminaco A.G. vs. Controller of Patents* Decision in 2002, in ways reminiscent of the infamous 1980 *Chakrabarty* case in the US, the Calcutta High Court overturned a long-standing policy of the Indian Patent Office by ruling that processes used to produce living organisms (e.g. vaccines containing live viruses) were eligible for protection under India's 1970 Patent Act (Basheer 2005). This decision signalled a profound break with established legal, cultural and religious norms in India, a country where 'life patenting' remains unpalatable to so many (Ramakrishna 2002).

This was not the only signal of the significant changes to the Indian IP regime that had been in the making since the late 1980s. The same year of the *Dimminaco* ruling—and overshadowing somewhat its importance—the 2002 amendment of the Patent Act was enacted that, among other things, brought microorganisms, as well as biochemical, biotechnological and microbiological processes, within the remit of patentable matter. Like other IP draft bills emerging from the government, it too was fiercely opposed by a number of civil society groups and representative of the generics industry. Cooperating with the Working Group on Patents and the Left parties in Lok Sabha, their concerns about the radical expansion of patentable matter took the form of two concrete demands: the narrowing of patentability to new chemical entities only and limiting the definition of microorganisms, taking advantage of the fact that TRIPS does not define either. These demands were cunningly manoeuvred through striking a compromise with the Left parties, whereby their vote in support of the government's bill was secured in exchange for the promise that an expert group would consider these two proposals *after* the final 2005 bill was passed (Mueller 2007). An expert group was established, but it was headed by Mashelkar, who, as we have seen, had long championed a 'robust' patent regime; unsurprisingly, the group rejected both proposals as incompatible with TRIPS in 2007 and again in 2009.[10] The 2009 report was promptly accepted by the government (Basheer 2009).

Although it cited the obligation imposed by TRIPS Article 27.3b to justify its rejection of narrower patentability criteria related to biotechnology, the reasons for the position of the Indian state are more likely to be found in its vision of propelling the pharmaceutical sector on a more innovative path and of creating a successful biotech sector. Having been singled out as priority sector since 1980, the view that private IP were indispensable to the development of the biotech sector had become dominant in key governmental circles by the late 1990s (Scoones 2002; Rajan 2006; Mueller 2007). It is revealing that as its 2002 patent bill draft was prepared for domestic consumption along the lines mentioned above, the Indian state was actively involved in international discussions over the same Article 27.3b as part of the TRIPS built-in review process that had started at the WTO in 1999. While the Indian state insisted, alongside other developing countries, that the review was not merely procedural but had to focus on substantive issues, including those related to the ethics and consequences of patenting life forms, unlike many of them, it was not insisting on a global ban on 'life patenting'.[11] But as the TRIPS deadline was still in the distance, many developing countries had not yet changed their patent laws to accommodate biotech patenting; in other words, 'life patenting' was by no means an established international norm in the late 1990s. For its part, the Indian state would dedicate immense energy at the WTO TRIPS Council and at the WIPO from then onwards to the issue of TRIPS compatibility with the CBD, especially on the issue of utilisation and *patenting* of genetic resources, effectively undermining any substantial revision of Article 27.3b or a potential ban on 'life patenting'. Since the early 2000s, the Indian state became the main proponent of reconciling TRIPS and the CBD through amending Article 29 of TRIPS to include mandatory disclosure of origin as criteria for the patentability of innovations based on genetic resources, an idea that had been floated in domestic debates over the Indian Biodiversity Bill during the late 1990s (Anuradha et al. 2001).

Nearly a decade earlier, the CBD negotiations had offered the Indian state an opportunity to pursue a simultaneously assertive and defensive strategy: to deal with the issue of biodiversity, which—as Indian environmental groups had been arguing at least since the early 1980s—required protection both from the negative effects of industrialisation and from misappropriation in the form of neocolonialism/biopiracy (Bose 2010; Krishnaswamy 2011), and to pursue more assertive goals. These included demands that access to genetic resources be made conditional upon the transfer of technologies developed by biotech companies in frontier economies, in line with the state's aim of developing an indigenous biotech sector in India (Rajan 1994). In the wake of its failure to achieve the latter goal—but, importantly, having succeeded in getting the principle of national sovereignty over genetic resources enshrined in the CBD and no burdensome obligations to preserve biodiversity—the state's offensive–defensive strategy was modified in line with the competitive knowledge economy imaginary that was taking hold during the mid-1990s. Seeking to make India a competitive economy was itself a partial manifestation of this two-pronged strategy; with regard to genetic resources, simultaneously to resist acts of biopiracy fuelled by a nationalist indignation against neocolonial expropriation and to become a competitive player in global biotech markets itself (Rajan 2006).

Many of the IP-related measures undertaken during this period regarding genetic resources bear the marks of this contradictory and ambivalent attitude. The Indian state moved relatively fast to defend the genetic resources under its control: in two highly publicised cases, it challenged patents issued by the US Patent Office on turmeric and basmati rice and succeeded in their revocation or limitation in 1997 and 2001, respectively, a move that, besides boosting its anti-(neo)colonial credentials at home, also strengthened the case for more 'robust' IP laws in India so as to protect *Indian* knowledge and genetic resources; the 2002 patent amendment discussed above prohibited the patenting of traditional knowledge, including that related to the properties and traditional uses of local genetic resources; the same amendment requested patent applicants to disclose the origin of genetic resources relevant to the application, making India among the earliest adopters of this means of linking IP and biodiversity laws; the 2002 Biodiversity Act, enacting the CBD, set up a system to police access to genetic material, at the helm of which stands a central state agency (the National Biodiversity Board); an initiative started by a number of NGOs to safeguard local communities' rights via the creation of People's Biodiversity Registers during the 1990s was co-opted in the 2002 Biodiversity Act as a continuous country-wide exercise, accompanying the state's own Traditional Knowledge Digital Library established in 2001 and currently running to over 30 million pages; following these measures, the Indian state negotiated deals with foreign patent offices, granting them access to such records for examination purposes, while playing an active role in international contests demanding that TRIPS and the CBD be linked via an international and legally binding disclosure of origin regime for genetic resources.

At the same time, these measures reveal another and rather more assertive tendency: in collecting and digitising vast amounts of India's public and traditional knowledge in registers that have no clear legal status domestically or globally, the Indian state moved one step closer to enabling the enclosure and commodification of this knowledge and the genetic resources related to it, the condition for which were created with the extension of sovereign property rights—alongside private ones—over them in the CBD. While the latter (vaguely) recognised the rights of indigenous and local communities, India's Biodiversity Act (2002) failed to properly recognise such rights. Despite going through a decade-long process of consultation and activism on the part of civil society, and although the three-tier structure enshrined in the Act gave local communities a role to play in biodiversity management, control over genetic resources and access to them have effectively been centralised under the National Biodiversity Board, an agency of a state with a poor record of protecting the rights of traditional/indigenous communities (Randeria 2007). Of over the 300 cases approved by the Board by the late 2000s, no cases of benefit-sharing with local communities had been concluded, nor had these communities been properly involved in the approval process (Kohli et al. 2009). Regardless, when India ratified the Nagoya Protocol of the CBD in 2012, it boasted that it had 'a great deal to offer to the world in terms of ... protecting biodiversity with the active participation of the local communities' (Jayanthi Natarajan, quoted in Suchitra 2012).

Although traditional knowledge is non-patentable, many such patents have been granted in India. In addition to those granted to Ayurvedic drugs mentioned earlier, a considerable number of public laboratories, especially CSIR ones, have been granted patents on innovations based on genetic resources traditionally used in India. The Biodiversity Act itself only polices access to genetic material on the part of foreigners, suggesting that the problem being addressed is not the appropriation and commodification of genetic resources per se, but rather its beneficiaries (Krishnaswamy 2011). For instance, while the CSIR was challenging the turmeric patent granted to two pharmacologists of the University of Mississippi by the US Patent Office, it itself held patents at the same office on turmeric uses, as did other Indian companies (Schuler 2004). Neither turmeric nor basmati rice growers in India saw any benefits from the revocation of the patent claims in question, while the commodification of these and other genetic resources through patenting continues, not least by the Indian state itself. While the majority of patents granted in India belong to foreign entities, the most prolific domestic patent-holders are not Indian private companies, but rather government departments, state-owned companies and public research laboratories, most notably the CSIR (Rajan 2005; Dutz 2007). Patenting is not the only means of appropriating social/collective knowledge. The Geographical Indicators (GI) Act, which came into force in India the same year as the Biodiversity Act and officially justified to prevent the misappropriation of well-known products developed over time by local communities (e.g. basmati, Darjeeling tea), provided another avenue for such appropriation; by the end of the 2000s, most of the nearly 130 GIs issued were registered by a state department or agency (Krishnaswamy 2011).

In its role of negotiating the space between the national and the global, the Indian state was not doing the minimum required to meet certain international obligations related to genetic resources. On the contrary, given its orientation towards an innovation-led growth regime that it placed at the heart of India becoming a globally competitive economy, it not only played a key role in setting the conditions for the emergence of a modern biotech sector—including building a 'robust' IP regime, the sine qua non of a successful Indian biotech sector (GoI 2007)—but also became a key player in the domestic and international biotech markets itself. Unlike the generic pharmaceutical sector, which was well established by the time the Indian state adopted a more competitive orientation, the biotech sector was in its infancy at best. Today, the Indian biotech sector is said to be booming, growing at a rate of 15–20% a year, dominated by private companies and boasting R&D levels that are relatively higher than the traditional pharmaceutical sector (Kumar et al. 2004). Despite these optimistic narratives, the sector's largest contributor is the biopharmaceutical sector (focusing on vaccines, biosimilars, stem cells etc.), indicating a strategy of building upon the strengths of the pharmaceutical sector rather than expanding on new areas such as agro-biotech. The biopharmaceutical sector, together with the bio-services sector, represents well over 80% of India's biotech sector's revenues (Sammut and Levine 2014: 7), but, importantly, neither is noted for its innovative capacities, focusing mainly on contract research, contract manufacturing and outsourced clinical trials, as mentioned above.

As in the US, the state has played a key role in the emergence of the Indian biotech sector: for example, investing early in human and physical infrastructure, financing direct R&D in public labs and universities, creating technology and science parks, and offering tax breaks and other measures to push businesses and public scientists to innovate, commercialise and patent (Mueller 2008). Moreover, taking its cues from institutional arrangements in the US, the Indian state took measures to create a venture capital market for high and risky technologies such as biotech since the late 1980s, financed, as in the US, initially with public funds (e.g. the Industrial Development Bank of India and State Finance Corporations) (Kumar et al. 2004; Rajan 2005). In tandem with state-financed programmes such as the Science and Technology Entrepreneurship Parks in the 1990s (now Technology Business Incubators), Small Business Innovation Research Initiative (2005) and Funds for Accelerating Start-Ups in Technology (2008), the liberalisation of venture capital rules from the mid-1990s onwards opened the gates for private venture capital funds entering the Indian biotech sector. The entry of private capital was helped in no small measure by the success of some Indian companies supported by (public) venture capital funds set up earlier and the unwavering commitment on the part of the state for the sector, boosted further by its embrace of population genetics (genomics) agenda around the same time (Rajan 2005). Despite the exhaustive measures taken to create a modern, competitive biotech sector in India, however, the case remains that the most aggressive players are public laboratories and state agencies; most private biotech companies and venture capital investors remain risk averse and—unlike in frontier economies—have so far preferred to invest in more traditional avenues, primarily in the manufacturing/production processes (Rajan 2006; Muller 2008). This orientation is symbolised by the trajectory of Biocon, an Indian biotech company established shortly after Genentech in the US. Although it reinvented itself along US biotech-inspired lines and its initial public offering in 2004 was as impressive as Genentech—only the second Indian company to pass the US$ 1 billion capitalisation mark on day one—it still focuses predominantly on production rather than on speculation in financial and/or IP markets (Rajan 2006).

Despite the Indian state's early commitment to developing an Indian biotech sector and subsequent investments, it was mainly after the dot. com bubble burst—but well before the performance of the sector could support it—that a powerful narrative about biotech's contributions to India's economic growth and entry into the global knowledge economy took hold (Scoones 2002; Dutfield 2004). Among the diverse and conflicting claims related to biotechnology stemming from scientific committees, populist movements, science policy analysts, peasant struggles, NGOs and the media, all relying on different versions of nationalism that often colluded with each other (Visvanathan and Parmar 2002; Cederlöf and Sivaramakrishnan 2006), the Indian state was steadfast in supporting the development of biotech as a priority for the nation. Slogans such as 'From IT to BT' and 'From the Green to the Gene Revolution'—depicting biotech as the next IT sector, an iconic symbol of India's global economic success, and as the second Green Revolution—struck various chords simultaneously (Scoones 2002; Newell 2003), although not always as intended. Narratives of agro-biotechnology providing food for the poor and transforming the ailing agricultural sector, in part inspired by the 'Malthus card' that proprietary seed and agrochemical companies adopted after their efforts to penetrate European markets failed, found early support in some prominent public figures associated with the Green Revolution in India, but also met with a strong, if fractious and conflicted, anti-GM campaign (Visvanathan and Parmar 2002; Scoones 2005; Ramanna 2006a). The campaign, featuring a number of vocal NGOs, such as Vandana Shiva's RFSTE (Research Foundation for Science, Technology and Ecology), Gene Campaign, SRISTI (Society for Research and Initiatives for Sustainable Technologies) and KRRS (Karnataka Rajya Ryota Sangha), reached wide proportions in 1998 over news of Monsanto's 'terminator' gene technology designed to create sterile seeds at harvest. Nevertheless, despite some successes with citizen juries, legal challenges in courts and even the burning of GM trial fields, no close alliances between NGOs and political groups on the GM issue emerged as compared with that on generic drugs (Newell 2003; Ramanna 2006a).

Although agro-biotech was and remains controversial in India, only six years after the first marketed GM crop appeared in the world market, the Indian government approved the first GM seed (Bt cotton) developed by Monsanto and the Maharashtra Hybrid Seed Company in 2002, signalling India's entry in agro-biotech markets, one of the first developing countries to do so. It is just as difficult to accurately estimate as it is to overestimate the impact that developments related to plant biotechnology in China—the catch-up contender and benchmark against which India's economic performance is often measured—had on India's policymakers. By the time the Indian government was contemplating the introduction of GM seeds in the late 1990s, the Chinese government was already investing annually five times the total of public and private investment on plant biotechnology in India (Huang et al. 2002: 675). But despite having singled out the development of an indigenous seed and agricultural biotech sector a key priority since the mid-1980s and having proceeded to invest considerable amounts of funds towards this end, China has been a cautious adopter of GM cultivation. Although it is a large importer of GM produce, it has permitted the commercial cultivation of only three GM crops—Bt cotton (1996), a transgenic poplar tree (2003) and a virus-resistant papaya (2006)—and is well behind the US, Brazil, India and other countries in terms of GM commercial cultivation (Wong and Chan 2016).

Nevertheless, the early commercialisation of Bt cotton in China and evidence of large amounts of public investment on biotechnology gave the Indian policymakers further impetus. Moreover, their hand was strengthened by the discovery of unlicensed Bt cotton fields in Gujarat just before the approval, an event that was seized by biotech enthusiasts within and outside the government as irrefutable evidence of farmers' free choice, weakening anti-GM groups' claims to be speaking on their behalf (Ramanna 2006a). Although the position of millions of Indian farmers vis-à-vis agro-biotech is impossible to ascertain—but is regularly preyed upon by groups with opposing views—the structural conundrum agriculture poses to the Indian state can be grasped in broad lines. Despite imaginaries of *India Inc.*, India is a predominantly agrarian country. Agriculture was relatively neglected by the post-independence state, not for lack of understanding of the reforms needed, but because its goal of combining capitalist accumulation with legitimation required the avoidance of political conflicts that would inevitably emerge if it intervened in the agricultural sector (Chatterjee 1993; Chibber 2012). Its most important intervention, the embrace of and institutional support for the Green Revolution, certainly required the state to stabilise and repress the social discontent that erupted at the time (Patel 2013). The Green Revolution left on its wake a sector that is in part industrialised and in others feudal, demonstrated partly by the very large and still growing number of small and marginal farms, currently accounting for around 80% of total holdings (Rao 2004; GoI 2014). Thus, when agro-biotech entered the fray in the late 1990s, the Indian agricultural sector was worlds apart from that of the frontier economies; although it generated a steadily declining share of the GDP, this was still sizeable, and, more importantly, it engaged anywhere between 65–70% of the total population—over 800 million people—most of whom live below the poverty line and rely on farming as the only source of livelihood (Rao 2004; Ramanna 2006b; GoI 2014).

Praise was heaped on the India state in 2001 for its pioneering role in recognising farmers' rights—absent in the CBD and non-binding in various international treaties—in its domestic law. Those praising made no reference to the manner in which the Indian state, when it challenged the turmeric and basmati patents, as discussed earlier, did so only on account of issues of interest to the Indian exporters and not of the multitudinous farmers dependent on such crops, nor of the manner in which the outcomes of a highly participatory process leading up to the 2002 Biodiversity Act, highlighting, among other things, the significance of biological diversity for the sustenance of the livelihoods of the vast majority of the population, were ignored and eventually rejected (Randeria 2007). The Indian Biodiversity Act, like the CBD, is silent on farmers' rights. Instead, praise was earned on account of the 2001 Protection of Plant Varieties and Farmers' Rights Act, which included, among others, the right of farmers to gain property rights over their own plant varieties alongside commercial breeders, as well as the right to save and exchange seed, undoubtedly an important right since around 80% of seeds in use are still exchanged in this traditional way in India (Ramanna 2006b).

There hardly existed any pressure on the part of Indian farmers for (IP) property rights over their plant varieties, and the right to save/exchange seed only became relevant once the Indian state embarked upon implementing TRIPS, which requires states to set up a *sui generis* system for protecting the rights of commercial breeders (PBRs). In practice, and not without considerable pressure from special interests, a *sui generis* regime has often taken a UPOV-like form, whose recent version not only does not recognise farmers' rights, but also significantly limits what they can do with proprietary seeds. Well before the 2005 TRIPS deadline, a plant varieties bill started to be drawn up in 1993; it only passed in 2001, following long and heated contests involving groups with divergent and conflicting interests (Ramakrishna 2002; Rao 2004). Although the NGOs involved had no unified position regarding farmers' rights, it is largely due to their mobilisation that these rights were recognised in the 2001 Act as a sort of compromise for accepting PBRs fashioned on the UPOV. The two most prominent views among civil society groups involved coalesced around farmers' rights as a form of IP right or as a developmental right; on the other side of the spectrum stood a relatively well-organised seed industry that appeared to have understood that the concept of farmers' rights as IP would reinforce their position and enable them to gain PBRs in India (Ramakrishna 2002; Ramanna 2006b; Ranjan 2009).

As in frontier economies earlier, agricultural plant breeding in India has been a public-sector stronghold; following India's embrace of a more competitive orientation, a large public sector involved in many functions, including breeding, research, production and extension services, at the core of which was the Indian Council for Agricultural Research and a number of excellent State Agricultural Universities, was made to open way for private plant breeders through a series of policy measures in 1985, 1988 and 1991 (Rao 2004; Ramanna 2006b). As their presence in the seed market grew from the early 1990s onwards, commercial seed companies were certainly keen on gaining protection for their varieties (Scoones 2002). But representatives of the public breeding sector, too, appear to have lent their support to such protection (Ramanna 2006b). Although commercial breeders have tended to focus on the most profitable crops, public breeders remain important despite (or because of) their focus on high-volume, low-value seeds (Rao 2004). Speaking on behalf of both sets of breeders, the then Agriculture Minister argued repeatedly that without proper plant varieties protection in place, cultivars developed by private and public agencies would be freely available to foreigners for exploitation (Ranjan 2009). Based on this logic, only a year after the 2001 Plant Varieties Act, the Indian government *re*started its process of acceding to the UPOV—a process initiated in 1998—so as to guarantee international recognition of Indian plant varieties (Ramakrishna 2002; Ranjan 2009).

Concerned that UPOV membership would require eliminating farmers' rights recognised in the 2001 Act, the (re)mobilisation of civil society groups reached a peak with the filing of a public interest litigation petition against the government in the Delhi High Court in October 2002 (Sahai 2004). Despite resistance at home, the Indian state is still in negotiations with the UPOV regarding its accession; negotiations are not public and both parties have so far maintained a studied silence on the matter. With the fate of farmers' rights in India still to be decided, it is worth pointing out that in granting commercial breeders, public institutions and farmers the right to have their plant varieties protected, the 2001 Act created the illusion of parity between these very different actors. In line with the government's interest in the commercialisation of India's genetic resources, the Act essentially envisaged a property rights model that lures farmers in as equal participants, despite the deep asymmetries between them and commercial and public plant breeders (Ramakrishna 2002; Ramanna 2006b). Procedures relating to farmers' variety registration are likely to be prohibitive for the average Indian farmer, and the benefit-sharing envisaged between commercial breeders and communities that have contributed to the development of a particular variety presupposes a vigilant, resourceful and entrepreneurial community (Ranjan 2009). In reality, barring the important right to save/exchange seeds, farmers' rights are generally quite difficult to uphold and unlikely to bring farmers any significant economic returns.

In conclusion, as the analysis above has shown, in attempting to make India a competitive player in the global knowledge economy, the Indian state can hardly be praised for an innovative approach itself, following instead the spirit—if not form—of the institutional innovations and arrangements set in place in frontier economies, particularly the US. But while the imaginaries of becoming a competitive market player and a global knowledge powerhouse are clearly adopted from the heights the Indian state seeks to reach, in practice, it has developed an ambivalent and ambiguous position underpinned by a ferociously nationalist drive to make India great but along lines and rules already (over)determined by those actors operating at the economic frontier. Hampered on the one hand by low education, health and other human development issues it has neglected for a long time, and propelled forward by a strong techno-scientific drive carefully nourished on the other, India is not yet a knowledge superpower. As we have seen, despite comprehensive efforts on the part of the state to make them globally competitive, including playing a significant role as a market player itself, the most dynamic parts of the Indian pharmaceutical and biotech sectors have so far been primarily orientated towards servicing foreign companies and wealthy markets. India's pharmaceutical and biotech sectors are some way off competing with those in frontier economies and still further, should the current orientation continue, from attending to the many needs of the Indian population. If the urgency of shifting attention to innovative and dynamic sectors that address these needs rather than compete in the IP race had caused a rethink of state strategy, this is not immediately apparent in the 2016 national IP policy, whose goal of bringing about *Creative India, Innovative India* demands even more social mobilisation: as an earlier draft revealed, '[t]he idea of being a creator and innovator must capture the imagination of our people to maximize the generation of all genres of IP rights' (Sridevan et al. 2014: 1).

Notes

1. Not only an essential activity of the state but a constitutional duty of the citizens; the Indian constitution may be the only one in the world that lists under citizens' duty that of developing a 'scientific temper' (Article 51A (h)).
2. For the early anti-colonial elite and especially the Indian Congress, independence was an opportunity to establish India as a presence on the world stage; in the words of Nehru, "A free India, with her vast resources, can be a great service to the world and to humanity. India will always make a difference to the world; fate has marked us for big things" (1939, quoted in Nayar 1979: 123).
3. CII is the Confederation of Indian Industry.
4. FICCI is the Federation of Indian Chambers of Commerce and Industry.
5. This goal was in fact formulated officially in the S&T policy of 2003; the fact that one of the authors of the Department of Science and Technology study that contributed to the policy, A.P.J. Abdul Kalam, became India's President (2002–2007) would have helped strengthen commitment to it, as well as to innovation and global competitiveness, on which it pivoted.
6. Despite healthy growth rates more recently, total public healthcare expenditure remains rather low in India, at around 1% of the GDP (Reddy et al. 2011: 8).
7. An executive act that has the same effect as that of an act of parliament but that needs to be approved within six weeks of the next parliamentary session, lest it lapses.
8. TRIPS 'mailbox' provision froze the novelty requirement for the granting of pharmaceutical patents for products which were filed after the TRIPS Agreement came into force (1995), but which would not benefit from patent protection in countries where this protection is not available until a later date (2005 for most developing countries). For example, if a pharmaceutical company filed a patent application in the US in 1996, it could also file a patent application through mailbox in India. When India started processing applications for pharmaceutical patents in 2005, examination of mailbox applications made between 1995 and 2005 was to be done not through examining 'prior art' as of 2005 but by reference to the prior art at the time of application (e.g. 1996). In addition to providing for mailbox application, the TRIPS Agreement also obliged states that do not need to comply with TRIPS until a later date to provide five years' exclusive market

exclusivity (EMRs) for products for which a mailbox application is made and which has obtained marketing approval in another country (the US in our example).

9. The 1962 war with China, while humiliating, helped to delineate India's aim of becoming a world power in more concrete terms through promoting prestige weaponry, power projection capabilities and 'big science'. Later, China's rise made it simultaneously a threat, a competitor and a benchmark against which India's own economic performance came to be measured (Alamgir 2009).
10. The furore caused by evidence that the 2007 report had plagiarised recommendations from another study, incidentally funded by private IP-reliant businesses, led to the restart of the work, once more headed by Mashelkar. The 2009 report was referenced properly, but the recommendations were virtually the same. See coverage of the Mashelkar report debacle here: http://lists.essential.org/pipermail/ip-health/2007-January/010407.html, last accessed 24 April 2016.
11. See for instance, Review of the Provisions of Article 27.3 (b), Communication from India at the WTO TRIPS Council, 29 October 1999, IP/C/W/161; see also communication from a group of countries, including India, at the same time: WT/GC/W/355.

Bibliography

Alamgir, J. 2009. *India's Open-Economy Policy: Globalism, Rivalry, Continuity.* Abingdon: Routledge.

Anuradha, R.V., B. Taneja, and A. Kothari. 2001. 'Experiences with biodiversity Policy-Making and Community Registers in India. In *Participation in Access and Benefit-Sharing Policy*, ed. K. Swiderska, 1–59. London: International Institute for Environment and Development (IIED).

Basheer, S. 2005. Policy Style Reasoning at the Indian Patent Office. *Intellectual Property Quarterly* 3: 309–323.

———. 2009. Encouraging Drug Innovation. *LiveMint*, August 27. Available at http://www.livemint.com/Opinion/v4H58mw6oSVDJOl2hKkWbK/Encouraging-drug-innovation.html. Accessed 5 July 2016.

Bergeron, S. 2004. *Fragments of Development: Nation, Gender, and the Space of Modernity.* Ann Arbor: University of Michigan Press.

Bhagwati, J. 1993. *India in Transition: Freeing the Economy.* Oxford: Clarendon Press.

Bose, I. 2010. *How Did the Indian Forest Rights Act 2006 Emerge?* IPPG Discussion Papers, No. 39. University of Manchester. Available at http://re.indiaenvironmentportal.org.in/files/forest%20rights%20act%202006-emerge.pdf. Accessed 15 July 2016.

Bound, K. 2007. *India, the Uneven Innovator*. London: Demos.

Cederlöf, G., and K. Sivaramakrishnan. 2006. *Ecological Nationalisms: Nature, Livelihoods, and Identities in South Asia*. Seattle, WA: University of Washington Press.

Chatterjee, Partha. 1993. *The Nation and Its Fragments*. Princeton, NJ: Princeton University Press.

Chibber, V. 2012. Organized Interests, Development Strategies, and Social Policies. In *Growth, Inequality, and Social Policy in India*, ed. R. Nagaraj, 168–192. London: Palgrave Macmillan.

Dasgupta, Bilal. 1999. Patent Lies and Latent Danger. *Economics and Political Weekly* 17: 976–993.

Drahos, Peter. 2010. *The Global Governance of Knowledge*. Cambridge: Cambridge University.

Dutfield, G. 2004. *Intellectual Property, Biogenetic Resources and Traditional Knowledge*. Abingdon: Earthscan.

Dutz, Mark, ed. 2007. *Unleashing India's Innovation: Toward Sustainable and Inclusive Growth*. Washington, DC: World Bank.

GoI. 2003. *Science and Technology Policy 2003*. Available at http://www.csir.res.in/external/heads/aboutcsir/Policy2003.htm. Accessed 11 July 2017.

———. 2007. *National Biotechnology Development Strategy*. New Delhi: Department of Biotechnology, Ministry of Science and Technology.

———. 2012. *Invitation of Views on the Draft National IPR Strategy*. New Delhi: Department of Industrial Policy and Promotion.

———. 2013. *Science*. New Delhi: Technology and Innovation Policy, Ministry of Science and Technology.

———. 2014. Agricultural Census: All Indian Report on Number and Area of Operational Holdings, 2010–11, Ministry of Agriculture. Available at http://agcensus.nic.in/document/agcensus2010/CompleteReport.pdf. Accessed 15 July 2016.

Greene, W. 2007. *The Emergence of India's Pharmaceutical Industry and Implications for the U.S. Generic Drug Market*. The US International Trade Commission, Working Paper No. 2007-05-A. Available at https://www.usitc.gov/publications/332/EC200705A.pdf. Accessed 10 July 2016.

Gupta, A.S. 2010. B K Keayla: A Personal Reminiscence. *Economic and Political Weekly* 45 (51): 25–26.

Herring, R.J. 1999. Embedded Particularism: India's Failed Developmental State. In *The Developmental State*, ed. M. Woo-Cummings, 306–344. Ithaca, NY: Cornell University Press.

Hobson, J. 1997. *The Wealth of States: A Comparative Sociology of International Economic and Political Change*. Cambridge: Cambridge University Press.

Huang, J., S. Rozelle, C. Pray, and Q. Wang. 2002. Plant Biotechnology in China. *Science* 295: 674–676.

IDMA. 2009. Patents. Available at http://www.idma-assn.org/patents.html. Accessed 5 July 2016.

Islam, N. 2010. Indigenous Medicine as Commodity. *Current Sociology* 58 (5): 777–798.

Iyer, K.V.R. 2002. The Doha-Delhi Odyssey. *Frontline* 19 (6): 16–29.

Jenkins, R. 1999. *Democratic Politics and Economic Reform in India*. Cambridge: Cambridge University Press.

———. 2003. Neo-Liberalism Encounters India's Indigenous Political Traditions. *Economy and Society* 32 (4): 584–610.

Joseph, K.J., and D. Abrol. 2009. Science, Technology and Innovation Policies in India: Achievements and Limits. In *BRICS and Development Alternatives: Innovation Systems and Policies*, ed. J.E. Cassiolato and V. Vitorino, 101–131. London: Anthem Press.

Kapczynski, A. 2009. Harmonization and Its Discontents: A Case Study of TRIPS Implementation in India's Pharmaceutical Sector. *California Law Review* 97: 1570–1649.

Kavani, S. 1999. R A Mashelkar-Catalyst for Change. *Business India*, June 28–July 11.

Keayla, B.K. 2005. Amended Patents Act: A Critique. *Combat Law* 4 (2).

Khilnani, S. 1997. *The Idea of India*. London: Hamish Hamilton.

Kohli, A. 2006. Politics of Economic Growth in India, Part I & II. *Economic and Political Weekly* 6: 1361–1370.

Kohli, A. 2012. State and Redistributive Development in India. In *Economic Growth, Inequality and Social Development in India: Is Inclusive Growth Possible?* ed. R. Nagaraj, 194–226. Basingstoke: Palgrave Macmillan.

Kohli, K., M. Fareedi, and S. Shalini Bhutani. 2009. *Six Years of the Biological Diversity Act in India*. Kalpavriksh and GRAIN, Delhi/Pune. Available at https://www.grain.org/article/entries/4172-6-years-of-the-biological-diversity-act-in-india. Accessed 15 July 2016.

Krishnaswamy, S. 2011. Access to Knowledge and Traditional Knowledge Protection. In *Access to Knowledge in India*, ed. Ramesh Subramanian and Lea Shaver. Gordonsville, VA: Bloomsbury Publishing.

Krishna, V.V. 1997. A Portrait of the Scientific Community in India: Historical Growth and Contemporary Problems. In *Scientific Communities in the Developing World*, ed. J. Gaillard, V.V. Krishna, and R. Waast, 236–280. New Delhi: Sage Publications.

———. 2013. India. In *BRICS National Systems of Innovation: The Role of the State*, ed. M. Scerri and H.M.M. Lastres, 138–187. New Delhi: Routledge.

Kumar, N., U. Quach, et al. 2004. Indian Biotechnology – Rapidly Evolving and Industry Led. *Nature Biotechnology* 22 (12): 31–36.

Mahtaney, P. 2007. *India, China and Globalization*. Basingstoke: Palgrave Macmillan.

Matthews, D. 2011. *Intellectual Property, Human Rights and Development: The Role of NGOs and Social Movements.* Cheltenham: Edward Elgar.

Mueller, J. 2007. The Tiger Awakens: the Tumultuous Transformation of India's Patent System and the Rise of Indian Pharmaceutical Innovation. *University of Pittsburgh Law Review* 68: 491–641.

———. 2008. *Biotechnology Patenting in India: Will Bio-Generics Lead a Sunrise Industry to Bio-Innovation?* Legal Studies Research Paper Series, University of Pittsburgh School of Law, Working Paper No. 2008-2.

Mukherji, R. 2012. Ideas, Interests and the Tipping Point: Economic Change in India. *Review of International Political Economy* 20 (2): 363–389.

Muzaka, V. 2011. *The Politics of Intellectual Property Rights and Access to Medicines.* Basingstoke: Palgrave Macmillan.

———. 2015. The State as Facilitator and Legitimator of 'New' Capital Accumulation: The Case of Patent Reform in India. *Journal of International Relations and Development* iFirst: 1–24.

Nayar, B.R. 1979. A World Role: The Dialectics of Purpose and Power. In *India: A Rising Middle Power*, ed. John W. Mellor. Boulder, CO: Westview Press.

———. 2000. The Limits of Economic Nationalism in India: Economic Reforms under the BJP-Led Government. *Asian Survey* 40 (5): 792–815.

Newell, P. 2003. *Biotech Firms, Biotech Politics: Negotiating GMOs in India.* Institute of Development Studies, Working Paper 201, Brighton, UK.

Park, C., and A. Jayadev. 2011. Access to Medicines in India: A Review of Recent Concerns. In *Access to Knowledge in India: New Research on Intellectual Property, Innovation and Development*, ed. L. Shaver and R. Subramanian, 78–108. London: Bloomsbury.

Patel, R. 2013. The Long Green Revolution. *The Journal of Peasant Studies* 40 (1): 1–63.

Pordié, L., and J.-P. GAudillière. 2014. The Reformulation Regime in Drug Discovery: Revisiting Polyherbals and Property Rights in the Ayurvedic Industry. *East Asian Science, Technology and Science* 8: 57–79.

Prasad, A. 2009. Capitalizing Disease: Biopolitics of Drug Trials in India. *Theory, Culture, Society* 26 (5): 1–29.

Raghavan, C. 2003. Does India Have a Trade Policy? *The Hindu*, April 28.

Rajan, K.S. 2005. Subjects of Speculation: Emergent Life Sciences and Market Logics in the United States and India. *American Anthropologist* 107 (1): 19–30.

———. 2006. *Biocapital: The Constitution of Postgenomic Life.* Durham, NC: Duke University Press.

———. 2007. Biocapital as an Emergent Form of Life: Speculations on the Figure of the Experimental Subject. In *Biosocialities, Genetics and the Social Sciences: Making Biologies and Identities*, ed. S. Gibbon and C. Novas, 157–186. London: Routledge.

Rajan, M.G. 1994. *India and the North-South Politics of Global Environmental Issues: The Cases of Ozone Depletion, Climate Change and Loss of Biodiversity*. Unpublished Ph.D. Thesis, Oxford University, Oxford.

Ramakrishna, T. 2002. *Development of IPR Regime in India with Reference to Agricultural Biotechnology*. Centre for Intellectual Property Rights, Research and Advocacy (CIPRA), National Law School of India University, Karnataka. Available at https://www.ids.ac.uk/files/Ramakrishna_IPR.pdf. Accessed 10 July 2016.

Ramanathan, U. 1996. Displacement and the Law. *Economic and Political Weekly*, June 15, pp. 1486–1491.

———. 2011. Land Acquisition, Eminent Domain and the 2011 Bill. *Economic and Political Weekly* 46 (44–45): 10–14.

Ramanna, A. 2002. Policy Implications of India's Patent Reforms: Patent Applications in the Post-1995 Era. *Economic and Political Weekly* 37 (21): 2065–2075.

———. 2005. Shifts in India's Policy on IP: The Role of Ideas, Coercion and Changing Interests. In *Death of Patents*, ed. P. Drahos, 150–174. London: Lawtext Publishing.

———. 2006a. *India's Policy on Genetically Modified Crops*. Asia Research Centre, Working Paper 15. London: LSE.

———. 2006b. *Farmers' Rights in India: A Case Study*. Background Study 4, The Farmers Rights Project. Lysaker, Norway: The Fridtjof Nansen Institute.

Randeria, S. 2003. Glocalization of Law. *Current Sociology* 52 (3/4): 305–328.

———. 2007. The State of Globalization: Legal Plurality, Overlapping Sovereignties and Ambiguous Alliances between Civil Society and the Cunning State in India. *Theory, Culture, Society* 24 (1): 1–33.

Rangnekar, D. 2006. No Pills for Poor People? *Economic and Political Weekly*, 4 February: 409–417.

Ranjan, P. 2009. Recent Developments in India's Plant Variety Protection, Seed Regulation and Linkages with UPOV's Proposed Membership. *The Journal of World Intellectual Property* 12 (3): 219–243.

Rao, C.M. 2004. Indian Seed System and Plant Variety Protection. *Economic and Political Weekly* 39 (8): 845–852.

Reddy, K.S., V. Patel, P. Jha, et al. 2011. Towards Achievement of Universal Health Care in India by 2020: A Call to Action. *Lancet* 377: 760–768.

Redwood, H. 1994. *New Horizons in India: The Consequences of Pharmaceutical Patent Protection*. Felixstowe: Oldwicks Press.

Reichman, J.H. 1998. 'Securing Compliance with the TRIPs Agreement after US vs. India. *Journal of International Economic Law* 1 (4): 585–601.

Rodrik, D., and A. Subramanian. 2004. *From 'Hindu Growth' to Productivity Surge: The Mystery of the Indian Growth Transition*. NBER Working Paper 10376.

Rudolph, L., and S. Rudolph. 1987. *In Pursuit of Lakshmi*. Chicago: University of Chicago Press.

Sahai, S. 2004. India Needs a Biotechnology Policy. *Frontline* 21 (10). Available at http://www.frontline.in/navigation/?type=static&page=flonnet&rdurl=fl2110/stories/20040521001708200.htm. Accessed 5 July 2016.

Sammut, S., and D. Levine. 2014. Accelerating Growth: Forging India's Bioeconomy. https://www.bio.org/sites/default/files/files/Burrill_AcceleratingGrowth_India-6-9-final.pdf. Accessed 19 June 2017.

Schuler, P. 2004. Biopiracy and Commercialization of Ethnobotanical Knowledge. In *Poor People's Knowledge*, ed. J.M. Finger and P. Schuler, 156–182. Washington, DC: World Bank.

Scoones, I. 2002. Biotech Science, Biotech Business: Current Challenges and Future Prospects. *Economic and Political Weekly* 37 (27): 2725–2733.

———. 2005. *Contentious Politics, Contentious Knowledges: Mobilising against GM Crops in India, South Africa and Brazil*. Institute of Development Studies, Working Paper 256, Brighton UK.

Shaver, L. 2011. Access to Knowledge: From Development to Human Rights. In *Access to Knowledge in India: New Research on Intellectual Property, Innovation and Development*, ed. L. Shaver and R. Subramanian, 1–13. London: Bloomsbury.

Sridevan, P., P.M. Singh, N. Sabharwal, et al. 2014. *Draft National IPR Policy*. Department of Industry Policy and Promotion. Available at http://dipp.nic.in/English/Schemes/Intellectual_Property_Rights/IPR_Policy_24December2014.pdf. Accessed 16 July 2016.

Suchitra, M. 2012. India Has Approved Ratification of Nagoya Protocol. *Down To Earth*, October 9. Available at http://www.downtoearth.org.in/news/india-has-approved-ratification-of-nagoya-protocol--39280. Accessed 25 January 2017.

Vielle, P. 1988. The State of the Periphery and Its Heritage. *Economy and Society* 17 (1): 52–89.

Visvanathan, S., and C. Parmar. 2002. A Biotechnology Story: Notes from India. *Economic and Political Weekly* 37 (27): 2714–2724.

Waldmeier, P. 2015. China Looks to Traditional Medicine as Tonic to Boost Growth. *Financial Times*, November 12. Available at https://www.ft.com/content/5b1969f0-7bce-11e5-a1fe-567b37f80b64. Accessed 2 April 2017.

Wilson, P., and A. Rao. 2012. India's Role in Global Health R&D' Centre for Global Health R&D Policy Assessment (Washington, DC, USA). Available at http://healthresearchpolicy.org/sites/healthresearchpolicy.org/files/assessments/files/R4D%20-%20Indias%20Role%20in%20Global%20Health%20RD%20Final.pdf. Accessed 10 July 2016.

Wong, A.Y., and A.W. Chan. 2016. Genetically Modified Foods in China and the United States: A Primer of Regulation and Intellectual Property Protection. *Food Science and Human Wellness* 5 (3): 124–140.

CHAPTER 6

Intellectual Property for Pharmaceuticals and Plant Genetic Resources in Brazil

The fervent project of nation-building of Vargas' *Estado Nôvo* (1930–1945) found an unlikely ally in the 'second discovery of Brazil' movement, a significant cultural movement that sought to couple the quest for modernity with the quest for *brasilidade* some 100 years after Brazil's independence (1822). Given its rebellious drive, it was a bitter irony that this modernist movement should contribute to the invention of an 'authentic' Brazilian national identity that colluded with that of the *Estado Nôvo*. One way such collusion symbolically manifested itself was the adoption of *Pau-Brasil*—the tree that gave Brazil its name—as the emblem of the movement and the concomitant enrolment of nature by the *Estado Nôvo* as the preferred locus of *brasilidade* (Philippou 2005). Perhaps this was not surprising, given the long-standing image of Brazil as an exuberant land of seductive natural beauty and an Edenic garden. Besides, the embedding of political imaginaries such as those of 'the nation' in nature has not been limited to Brazil. But more than elsewhere, it has been in post-colonial societies that different versions of the relationship to nature and the extent and form of control over it as resource and/or heritage have converged and become part of the processes of legitimising and consolidating a nation (Cederlöf and Sivaramakrishan 2006). It may be added that, in the case of Brazil, as nation-state-building efforts sought to obfuscate the existence of different peoples in a context purportedly 'lacking' of a past comparable to the one available to European nation-builders, appeals to nature were even more important than elsewhere.

V. Muzaka, *Food, Health and the Knowledge Economy*, Building a Sustainable Political Economy: SPERI Research & Policy, DOI 10.1057/978-1-137-59306-1_6

But the nature that became the locus of the new 'authentic' Brazilian identity was hardly paradisiacal; on the contrary, it was made and remade in the constant and conflictual encounters with the Old World and plugged into the world economy since the Iberian conquest. It is revealing that *Pau-Brasil*—the most valuable product the Portuguese initially took from Brazil on account of the coveted red dye extracted from it—was declared extinct in the wild in 1928, the moment in time when the movement that adopted it as the symbol of a Brazilian identity free from the economic and cultural dependency on the Old World gained speed. It reveals more still: its exploitation marks the beginning of the end of *Mata Atlântica*, the tropical forest that once covered the entire Brazilian coast and of which only a few small patches are left. Such deforestation paved the way for sugarcane, coffee and later soy cultures, alongside the associated local and global socio-economic changes they engendered. Arriving in Brazil through the circuitous routes of the imperial botanical chess game that locked the great colonial powers in constant economic and geopolitical conflict, these non-native plants would go on to provide more cause for national economic pride: Brazil becoming the largest or second-largest world exporter of sugar, coffee and, more recently, soy, an outcome accompanied by fresh waves of socio-economic transformations, conflicts and natural destruction.

Exploited as a resource and sometimes preserved as heritage, nature as an important locus of national identity would also come to support another dimension of nation-building characteristic of newly formed states in the Americas: a forceful appeal to the future, which first materialised in the US as the original 'land of the future' and which in Brazil was based, to a significant extent, on the belief that the country's size and natural wealth were a guarantee of its future political greatness (Carvalho 2000; Hurrell 2005). Once part of the official ideology and echoed in the national anthem, the yearning for national greatness and power status would be ruthlessly exploited by Brazil's military governments and is not an insignificant undercurrent in Brazilian politics today. The fusion of these themes can be observed in that most iconic of national and nationalist symbols: *Amazônia*. Drawn into economic and geopolitical conflicts and global markets since the 1500, it was during the 'second discovery of Brazil' that *Amazônia* became not only a repository of markers of national distinctiveness, but perhaps more importantly, in the nationalist discourse of the *Estado Nôvo*, it became the metonym of Brazil itself, the region of the future in the land of the future (Garfield 2013). As Vargas said in his 'Speech of the Amazon River' in 1940, 'In the same way that the image of [the Amazon River] is for Brazilians a measure of the greatness of Brazil, your problems ... are those of the entire nation' (quoted in Garfield 2013: 47), that is, a land of untold potential but blighted by social injustice, of natural bounty desecrated by plunder, condemned by history but soon to be liberated by science, deformed by open markets but in the process of being restored by the state. From then on, and especially during the military period (1964–1985), when the state actively promoted *A Amazônia é Nossa!* (Amazon is Ours) as a propaganda tool of national pride and sovereignty, the nationalisation of the Amazon 'question' led to formidable state activity in the form of infrastructure investment, the luring in of private (especially foreign) capital for mega-projects and resources extraction, logging, clearing, expanding pasture lands and so on. This was a process that, under the guise of national development, had destroyed more than 10% of the Brazilian Amazon by 1985 (Kolk 1996).

Neither this state of affairs, nor the end of the military regime appears to have reduced the appeal of the Amazon as a political symbol of Brazil's great power status and sovereignty; on the contrary, the more environmental concerns appeared in the global agenda, the more frequent its use.[1] Indeed, the simultaneous emergence of a global environmental consciousness and the touted promises of modern biotechnology only served to multiply claims over the Amazon (e.g. the treasure trove of biodiversity, the pharmacopeia of the world, the lungs of the world etc.) and strengthen its status as a symbol of national identity and fortune. A recent manifestation of these tendencies can be found in the somewhat belated Biotechnology Development Policy of the Brazilian federal state, launched in 2007. Having first portrayed his administration as the preserver of nature and defender of the Amazon, President 'Lula' da Silva announced the policy as 'another path to the future' in which Brazil 'will never again be a supplier of raw materials for the world market'; instead, harbouring vast forests and 20% of world's biodiversity warranted Brazil becoming 'the great technological and environmental power of the 21st century' (Lula 2007). Accordingly, the anticipated leadership position in biotech in a matter of 10–15 years was to be similar and closely related to the one that Brazil had already assumed in the world biofuels sector, a reference to Brazil's success in exploring the frontiers of science as much as those of its territory, which bestowed upon it the title of 'the natural knowledge economy' (Bound 2008).

This more recent economic and political imaginary of becoming a global knowledge power, embedded in nature in the case of Brazil, represents a local refraction of the knowledge economy imaginary that, as discussed, was a state project initially espoused by those advanced economies that had a competitive advantage in the new technologies of the 1970s and 1980s. Questions about Brazil's ability to achieve the feat of becoming a great knowledge power in the changed global context reverberate with another long-running concern at the core of its national identity: its imitative propensity. Uneasy about what one of the best Brazilian writers had described in the late 1800s as the tendency for a foreign impetus to determine the direction of movement,[2] many involved in Brazil's modernist movement of the first part of the twentieth century tried to reflect on, explain and purge this tendency, only for inauthenticity to become in time the most authentic element of the Brazilian identity (Schwartz 1988). This had nothing to do with the nature of the Brazilians, of course, and everything to do with the specific historical conditions that shaped the making of Brazil: an independent state created without a revolution, where the colonial socio-economic structures remained intact and whose oppressive forms of life merged gradually and only imperfectly with ideas of modernity and progress (Schwartz 1988). Celso Furtado, relying on an understanding of Brazil's socio-economic realities that remains unparalleled, captured the economic and political manifestation of this tendency in the concept of 'mimetic development' (1992, 1997). This was a model of industrialisation geared towards the satisfaction of the demand created by a small elite where wealth was concentrated and whose consumption patterns were similar to elites in frontier economies, a dynamic that over time led to a superficial modernisation of Brazil's socio-economic structures accompanied by and perpetuating two enduring and debilitating features of Brazil's political economy: income concentration and external indebtedness. It could be argued that as long as these fundamental problems persist, and despite institutional transformations undertaken in the areas of innovation, S&T and IP inspired by those designed in frontier economies, the goal of becoming a great technological power may remain some time off in the future.

6.1 On the Economic Imaginary of Brazil as the Natural Knowledge Power

Having started his academic career with a study on Brazil's racial history and consolidated his reputation as a dependency theorist, Fernando Henrique Cardoso (FHC), the president who would oversee Brazil's most recent and significant transformation, laid out in the 14th National Forum in 2002 Brazil's principal challenge thus: '... becoming competent, capable of being positively inserted in the worldwide economy. And positive insertion, today, means insertion with knowledge ... our future will depend on our capacity to advance systematically towards this goal' (FHC 2002: 21). FHC was not the first to recognise this challenge. Having created the Ministry of Science and Technology in 1985, which placed innovation on the policy agenda for the first time (Cassiolato and Schmitz 1992), the first civilian government had appointed in 1986 a special group to propose a new growth strategy that would eventually pronounce the import substitution industrialisation strategy to have exhausted the prospects for growth. Reflecting the strengths that ideas about the need for competitive restructuring were acquiring in light of significant changes in advanced economies—in networks around the BNDES,[3] the Society for the Advancement of Science and IEDI[4] in Brazil and CEPAL[5] in Chile, among others—it proposed a turn to structural competitiveness and the dynamic insertion of Brazil in the world economy (Marques 1999; Pedersen 2008; Leiva 2008). As in frontier economies, structural competitiveness required a new orientation of state activities. FHC, the self-professed articulator of the 'new Brazilian state', was clear that '[t]his new phase of capitalism does not necessarily limit states, it also opens up new perspectives for states' (FHC 2003, quoted in Biehl 2004: 114). However, while such positive insertion was portrayed unfailingly in every official communication to depend on changes in the productive structure towards higher levels of technological sophistication and innovation, the Brazilian state during this period presided over the opposite: the hollowing out of its productive base, the falling levels of technological intensity and the unravelling of existing learning and innovative networks and capabilities (Coutinho 2003; Lastres et al. 2003).

These tendencies accelerated especially under FHC's own watch. It is true that the capacities of the Brazilian state were not what they once had appeared to be. Having surprised the world with its 'miraculous' double-digit growth rates when most advanced economies came under the grip of recession only to be found heading the largest debtor countries' list in 1982, the Brazilian state's capacities weakened during the structural adjustment period that followed the debt crisis. This period saw deep cuts in public expenditure, no less than eight different plans for monetary stabilisation, the use of four different currencies, six wage freezes, 18 changes in foreign exchange rules and 22 proposals to renegotiate foreign debt, whose service continued to demand an ever-higher share of the nation's income (Das 1994: 416). But once some stability was achieved through the 1994 *Plano Real*, the worsening of Brazil's competitive position was less the outcome of a weakened state apparatus than of the adoption of an economic imaginary on the part of a core executive elite headed by FHC himself. Reflecting the neoliberal faith in markets—where technology and knowledge could be acquired like any other commodity, an assumption at odds with the strategies frontier states and high-tech firms were adopting to protect and extend their technological rents—it promoted Brazil's 'competitive insertion' primarily through the opening of the economy and FDI flows (Rocha 2002; Coutinho 2003; Leiva 2008). Lured by the privatisation, deregulation and liberalisation policies of the 1990s, FDI did flock to Brazil, raising its share to GDP dramatically from a historic average of 10% to around 25% by the late 1990s (Cassiolato et al. 2014: 80). But because the concomitant rise of neoliberalism and financialisation transforming global economic structures along the lines discussed earlier had unleashed not the Schumpeterian instinct among market players but rather the predatory-rentier one, most of the FDI simply went towards strengthening foreign control of existing assets in the economy. By the time FHC gave his 'Knowledge Society' speech to the 14th National Forum in 2002, neither a technologically intensive pattern of industrialisation, nor the 'competitive insertion' of Brazil in the global economy had occurred. On the contrary, the share of industry fell from 44% in 1980 to as low as 27% of the GDP by the late 1990s, and the share of high-tech manufactures represented only 7.9% of Brazil's total merchandise exports by the time FHC left office, betraying a return to a comparative advantage based on natural resources and manufactures based on them (Rodriguez 2008; Palma 2012).

Launching the *Avança Brasil* (Brazil Forward) plan soon after his second term began—a 'national project for a new Brazil' that strongly emphasised the importance of transforming Brazil's socio-economic structures through the application of science, technology and innovation but this time aware of the need to address 'market failures' that had become apparent during his first term—FHC had occasion to boast about Brazil's scientific and technological base as, barring frontier economies, 'the largest in the world' (FHC 2000). In Latin America, certainly. Benefitting from relatively large public investment from the early 1950s, a significant S&T base had been built by the 1980s, whose strength was most visible in a number of excellent public universities and research institutes (Schwartzman 1995; Koeller and Cassiolato 2009). As in India, investment in S&T was profoundly elitist in nature and accompanied by a comparative neglect of basic education. Despite more recent increases in public investment in education, Brazil remains the only country where public investment per student in higher education—incidentally, most likely to come from the fifth-highest income quintile—is on par with OECD levels, whereas those in basic and secondary education are around five times lower (in purchasing power parity [PPP] terms) (OECD 2001; Koeller and Gordon 2013). Two of the most visible consequences of this orientation—not unlike India—are that, first, success remained limited to a few high-tech 'pockets of excellence' (e.g. aeronautics, agriculture research and petrochemicals), and, second, public universities and research institutes, having been the main beneficiaries of S&T investment, would become key players in the knowledge economy project Brazil came to adopt later.

Following a similar pattern to the reorientation of the US S&T policy during the 1970s that demanded universities play an active role in re-establishing the US competitive position, the Brazilian state attempted during the late 1980s and 1990s to make research carried out in public institutions more directly relevant to its international competitiveness (Schwartzman 1995). This it tried without the facility of the US state to increase public investment in S&T, given that the debt crisis had not only turned off the spigots of cheap foreign borrowing for Brazil, but also because its T-bonds went towards servicing the ever-increasing debt burden. Up until that point, much like the Keynesian state in frontier economies, the Brazilian state had sought to harness knowledge's contribution to economic growth through considerable public investment in science, whose outcomes, likewise, were generally freely available. IP did not play a significant role during this period in Brazil even if the country had been one of the original signatories of the 1883 international Paris Convention (on patents). Patents' 'Dark Ages' started when, in two significant overhauls of Brazil's patent system in 1945 and 1971, a long list of products and processes (e.g. pharmaceutical processes and products, therapeutic techniques, microorganisms, chemical products etc.) were excluded from patentable subject matter (Mazzoleni and Póvoa 2009). In principle, these steps were taken with a view to promoting the development and technological upgrading of these sectors. Likewise, the patent office (INPI[6]) created in 1970 was given the task of overseeing technology transfer contracts between domestic and foreign companies in order to ensure that such transfer resulted in technology learning and upgrading. In international regimes at the time, as discussed earlier, the Brazilian state was also leading calls to reform the international IP regime to enhance technology transfer to developing countries. While this indicates consistency in IP orientation during this period, the same cannot be said about the developmentalist orientation of the Brazilian state.

Responding to a profound sense of lag and lack accompanying Brazil's self-image as an underdeveloped country in the 1930s, the Brazilian state took upon itself the task of engineering and achieving national development through industrialisation, the only route it saw to catch up with frontier economies. In the process of negotiating the contradictory space between the national and the global, although it rarely failed to use an apparent threat (e.g. by foreign companies, governments or international institutions) to its autonomy in carrying out this developmentalist task to rally nationalist sentiments, it had apparently little difficulty in not only allowing but often actively encouraging foreign capital to take over the most dynamic parts of the Brazilian economy. Having already stood at 41% of total industrial capital in 1940, foreign capital's share increased to nearly 70% in the auto sector and 62% in the pharmaceutical sector in 1960, for instance, increasing further after the military coup to 100% and 80%, respectively, in 1968 (Ackerman 1971: 21). Despite the Brazilian state's international stance regarding technology transfer and domestic IP and S&T policies during this period, its FDI policy pulled in the opposite direction, with consequences in its socio-economic fortunes still unfolding today. For our purposes, the key issue with such high control—which was to increase further during the de-nationalisation of the economy of the 1990s—was the lack of control over knowledge and technology because, having invested primarily to enter the domestic market, foreign companies occupied the most dynamic sectors without being particularly innovative, thus dampening the technological dynamism that might have been achieved otherwise (Kohli 2004; Cassiolato et al. 2014).

This phenomenon was exacerbated by another and rather more challenging problem. Before Brazil's fortunes would take a turn for the worse, under the guise of national development, the Brazilian state had overseen an economic 'miracle', better described as 'economic growth without development' (Furtado 1992) and a case of 'unaimed opulence' (Drèze and Sen 1989), that excluded the vast majority of the Brazilian population. Not only did the historically high-income concentration not improve as the economy grew, but it worsened: for instance, the income share of the lowest 20% of the economically active population fell from 3.9% in 1960 to 2.8% of the total in 1980 (Lamounier 1989: 134), nearly four times less than that in the much poorer India. Short-termism and ill-defined developmentalism had succeeded in building a significant industrial base by the early 1980s, but both consumer goods and capital goods sectors were largely directed towards satisfying a distorted demand structure, a dynamic that exacerbated the socio-economic marginalisation of large parts of the population (Furtado 1992, 1997). Such marginalisation did not put Brazil's economy in a good stead either, for the application of technology and the resulting degree of competitiveness of its industrial base varied greatly; high-tech sectors accounted for a relatively small share of manufactures and, as the ensuing liberalisation reforms would reveal, were generally unable to compete with the technologically complex manufactures from the Asian 'tigers', which had maintained low-income inequality levels, high investment rates, effective industrial policies and outstanding technology-absorbing capabilities (Rodriguez 2008; Feijo and Lamonica 2010; Palma 2012).

The worsening of most economic and social indicators during Brazil's 'lost decade' occurred when the knowledge economy imaginary was gaining strength in frontier economies and would come to orient the transformation of their institutional forms with a view to becoming competitive knowledge economies. This imaginary, in turn, held considerable appeal for the Brazilian elites, not only because the debt crisis had closed the door on the previous growth strategy, but also because they remained wedded to the view that frontier economies held the key to the only route to development. The implementation of this Western imaginary took a clearer shape in FHC's *Knowledge Society* project during his second term (1999–2002). The dedication to this project was visible in a number of measures, including various S&T institutional reforms inspired by those of frontier states; a new IP regime aimed at attracting more FDI but profoundly at odds with Brazil's productive, innovative and social structures; and, more importantly perhaps, macroeconomic policies implemented to enhance Brazil's positive insertion in the global economy but which ended up reinforcing the existing 'malign' macroeconomic context that undermined much of whatever potential these and other policies might have had.

S&T and innovation policies in particular were notable for their supply-side approach, in line with the Schumpeterian state orientation of most frontier economies. This included the controversial and often-reformed R&D tax incentives policy and the institutional innovation of S&T sectoral funds,[7] whose main merit so far has been that of providing consistency for public R&D funding (Rodriguez 2008; Koeller and Gordon 2013). More importantly for our purposes, this mechanism—still one of the main ones for supporting innovation in Brazil—was and remains part of an effort by the Brazilian state to use its significant scientific base to contribute to the goal of Brazil's positive insertion in the global knowledge economy. For instance, the goal of FINEP[8]—the most important public innovation agency in Brazil and in charge of the sectoral funds programme—was reorientated away from financing academic and basic research towards financing industrial technology research during the 1990s (Schwartzman 1995). Efforts on the part of the state to push public labs and universities towards a more active role in improving Brazil's economic fortunes, inspired by similar trends originating in the US, had started since the late 1970s (Koeller and Cassiolato 2009), but during this period, they were invigorated, even if the time ran out for the FHC administration to see through Brazil's version of the US 1980 Bayh–Dole Act,a task completed with as much conviction by the new Lula administration in the form of the 2004 Innovation Law. Likewise, targeted efforts were undertaken to create a venture capital market to support the spin-off companies that were expected to emerge during this period—for example, the authorisation of venture capital investment funds in 1994 and authorisations in 1994 and 1999 to enable public institutions and pensions funds to invest in venture capital vehicles—were accompanied by the channelling of public funds towards this end (Ribeiro and Carvalho 2008; Leamon and Lerner 2012). Efforts to create a venture capital market through considerable public investment, as had happened in the US, continued and expanded further during the new Lula administration, with BNDESPAR (created in 1982), the BNDES CRIATEC programme (created in 2007) and the FINEP INOVAR programme (created in 2000) becoming the most notable.

Simultaneously the clearest sign of Brazil's embrace of the knowledge economy imaginary and of its disarticulation vis-à-vis Brazil's socio-economic realities, the new 1996 IP law (Lei no. 9.279) represented a significant rupture with the past. Given the state of Brazil's economic structures and of its innovation system at the time, this shift could not be seen as an instance of the Brazilian state seeking to secure and protect technological rents on behalf of Brazilian capital. Instead, it is a manifestation of state's embrace of the knowledge economy imaginary and of the assumption that IP would help Brazil's new growth strategy, first, by attracting FDI flows and speeding up technological upgrading and, second, through an increase in time in patenting by Brazilian firms themselves.

It is true that, as discussed earlier, Brazil came under immense pressure from the US to change its position at the GATT since the beginning of TRIPS negotiations and would remain in US Special 301 list during most of the negotiations. This was no small matter, given the vulnerability of exports to US sanctions and, importantly, the strong ties of the export sector with members of the Brazilian Congress (Flynn 2011; Sweet 2013). Even so, it is notable that by 1994 the executive was fully on board and when FHC's first term began in 1995, the patent bill that had been debated in Congress since 1991 changed substantially; when it passed in 1996, it was not only considerably more generous to IP-holders than earlier drafts, but more so than TRIPS itself. Unsurprisingly, the relative share of Brazilian patent applications after the law took effect not only did not increase as was hoped, but it fell significantly compared with the pre-1996 period (Cassiolato et al. 2014: 97). The INPI, now handing out large numbers of state monopolies predominantly to foreign companies, shed its earlier reputation by redefining itself as part of a competitive knowledge economy in which Brazil was to become a winner (Drahos 2010).

As will be discussed shortly, the 1996 IP law did not resolve the tensions inherent in the new orientation of the Brazilian state; on the contrary, as they became more apparent, numerous changes to it would be proposed—many related to life science sectors—and some would be carried out. Starting from the late 1990s and accelerating during the Lula and PT (*Partido dos Trabalhadores*) periods, a move in two directions can be observed: on the one hand, the IP-maximalist approach was somewhat tempered, although by no means the drive to play the patenting race, and, on the other, the imaginary of Brazil as a great knowledge power became more dominant. The strengthening of this imaginary was helped by the rediscovery and restoration of industrial policy as a legitimate state tool towards achieving 'growth with equity', at least in principle. PITCE[9]—the first industrial policy launched soon after the PT came in power—made clear that the world economy was 'characterised by new economic dynamics ... that see innovation as the key element for national industrial and competitive growth' (PITCE 2003: 4). Since then, innovation became the *fil rouge* of all industrial and, obviously, S&T policies. This said, the S&T orientation remained largely supply side and did not differ radically from that of the earlier period, even if public funding increased substantially. All the important measures taken to support the industrial sector—for example, the creation of the Brazilian Industrial Development Agency in 2004, the multiplication of sectoral Competitiveness Forums, increased support for the National Program for Supporting Business Incubators and Science Parks, the launch of the Health–Industrial Complex in 2008 and all the industrial policies articulated since 2003, PITCE, PDP and *Brasil Maior*—emphasised without fail the importance of technological innovation in achieving 'genuine competitiveness'. The fortunes of the Brazilian agro-biotech and pharmaceutical sectors improved somewhat, as we shall see, but the transformation of Brazil into a knowledge power remains a project.

6.2 Reforming IP Rules for Pharmaceuticals

Distancing himself from the previous government that had insisted that the US Section 301 trade sanctions applied against Brazilian exports in 1988 were illegal and would not change Brazil's stance on pharmaceutical patents, presidential candidate Fernando Collor de Melo argued that the lack of pharmaceutical patent protection in Brazil was against its national interest (Shadlen 2017). Once installed in the *Planalto*, Collor sent a draft IP law to Congress in April 1991 that duly called for the immediate introduction of pharmaceutical patents. During most of the deliberations that concluded with its approval in 1996—and, as it would turn out, after its coming into force—pressure on Brazil from the US state and proprietary pharmaceutical companies remained intense. This pressure had much to do with the fact that, as discussed, Brazil was alone among middle-income countries that offered no pharmaceutical patent protection at all, a characteristic that, alongside its large domestic market and its vocal objection to the TRIPS agenda at the GATT/WTO, made it a clear target. It did not help that the Brazilian pharmaceutical market was already controlled by foreign pharmaceutical companies. Although the aim of eliminating pharmaceutical patent protection had been to promote a national pharmaceutical sector, the disarticulation of this policy from industrial policy—visible in measures such as Instruction 113 and the 'Law of Similars', which in practice offered strong incentives to foreign companies (Mazzoleni and Póvoa 2009)—led to the opposite outcome. Confounding latter-day proponents of the view that strong IP protection is a precondition for FDI flows, these actually increased the weaker IP protection in the sector, consolidating the share of foreign control from around 13% in 1930 to nearly 80% in 1969, a control sealed by the sixfold increase in FDI flows in this sector during the 1970s (Ackerman 1971; Gereffi 1983).

These developments had two substantial and perverse effects: the marginalisation of Brazilian pharmaceutical companies—both in size and in their focus on areas of no major interest to foreign firms—and the dampening of dynamism and R&D efforts in the sector, as foreign firms remained primarily focused on producing formulations and distributing imported drugs, and, in a handful of cases, on producing active pharmaceutical ingredients (APIs) (Evans 1979). Following the liberalisation reforms that started in the late 1980s and accelerated during the 1990s, imports of APIs and of finished goods increased and the national sector weakened further: only one among the top 20 pharmaceutical companies was nationally owned in the mid-1990s in Brazil (Flynn 2011; Caliari and Ruiz 2014). Public research institutes such as Fiocruz and Instituto Butantan, despite having accumulated significant research capacities since their creation at the start of the century, also lay largely dormant during the 1980s.

Although the structural position of the Brazilian national pharmaceutical sector was no match to the Indian one when it came to resisting changes to pharmaceutical IP rules, it was nevertheless actively involved in the intense contests over the new IP orientation the state was embarking on and which resulted in the 1996 law. This orientation occasioned strong societal action that pitched against each other groups with opposing aims. One group of actors, mobilised to secure the highest IP protection standards possible, included INTERFARMA,[10] an association of foreign pharmaceutical companies created in the late 1980s with this aim in mind, the Brazilian Association for Intellectual Property (ABPI), the main defender of IP in Brazil since the 1960s, ABRAPI,[11] a newly formed but not long-lasting association of the fledging Brazilian biotech sector, and the peak industry association, CNI[12]. The other consisted of a dense network of NGOs—including, among others, the Brazilian Confederation of Bishops, the Brazilian Academy of Science and the two national pharmaceutical sector associations, ABIFINA[13] and ALANAC[14]—united in opposition in the Forum for the Free Use of Knowledge (FLUC), which included nearly 1000 civil society organisations, alongside Left parties in Congress (Newell 2008; Shadlen 2017). Over the prolonged and fractious debates about the new IP law, which included no less than 1400 different amendments proposed, it was the executive that ultimately had the upper hand. Not unlike the Indian case, it favoured a strong IP law and manipulated concerns raised by export interests (e.g. paper and shoe manufacturers) about the potential loss of the (at the time rather important) US market to stroke fears about the grave consequences US reprisal would have for

Brazil's already weak economy (Flynn 2011; Sweet 2013; Shadlen 2017). Using a narrative that appeared to both alleviate the concerns of more traditional sectors and promise the arrival of a new era for Brazil's high-tech sectors of the future, the FHC administration was successful in getting the new law passed in 1996.

The new IP law was notable above all for its premature and onerous nature. Compared with the earlier drafts that had been discussed since 1991, the final version was decidedly more generous to IP-holders and, moreover, its generosity surpassed TRIPS' own. In addition to failing to take advantage of TRIPS flexibilities, for example, with regard to compulsory licensing provisions, parallel importing and research exception (Bolar) provisions,[15] the most significant 'TRIPS+' features were the criminalisation of patent infringements, the introduction of 'pipeline patents'[16] and the early implementation of pharmaceutical patent obligations, for which the deadline was, as for India, in 2005. In all these aspects, the Brazilian state was relinquishing the little policy space it itself, alongside other developing countries, had fought to secure during the TRIPS negotiations, but perhaps the most puzzling and damaging of them was its provisions for 'pipeline patents'. This retroactive extension of TRIPS patent obligations was vehemently opposed by Brazil and other developing countries during the negotiations, and by 1993, it was clear that TRIPS would make no demands for countries to acknowledge patents that had already been granted (but awaiting marketing approval) prior to 1995, much to the consternation of transnational pharmaceutical companies (Watal 2001; Sell 2003). But when the Brazilian executive became convinced of the necessity to provide foreign companies with an unmistakable sign of Brazil's commitment to patent rights in the form of 'pipeline patents', the companies' mood improved, at least temporarily: when the 'pipeline window' opened (May 1996–May 1997), 1182 'pipeline' applications were made—nearly half from US companies—and over 700 were granted, which in practice meant that at least 340 drugs would have lacked patent protection in Brazil post 1996 but for the pipeline mechanism (Sweet 2013).

If part of the rationale for these generous provisions had been that they would bring an end to the incessant pressure, it soon proved erroneous. Pressure continued and culminated with a WTO challenge by the US in 2000 related to 'local working' provisions in Brazil's new IP law (Article 68). The offending provision had been a fundamental precondition for the continued validity of a patent in the early international patent system and all industrial nations had required local working of patents with a view to promoting technological upgrading and national industrialisation (Gontijo 2005). Although the 'local working' provisions in Brazil's new IP law were effectively toothless and only very rarely used in their earlier and more robust form to override patent rights (Gosain 2007; Shadlen 2017), they were considered objectionable by the US pharmaceutical companies that persuaded the Clinton administration to take up the fight at the WTO. The case was never adjudicated, as the two parties reached a 'mutual understanding' according to which the Brazilian state had to notify the US were it contemplating using the provision against a US company, a clear manifestation of the lengths the US state would go to protect the technological rents of its companies and the former would oblige, for the Brazilian state never contemplated changing—let alone using—these provisions in the subsequent changes to the IP law.

Widely seen as humiliating in Brazil, the case had an unexpectedly positive effect: the mobilisation of civil society groups on the 'IP-access to medicines' front. Having been a significant part of the democratisation movement, *movimento sanitário* (the sanitary and healthcare movement) was already strong in Brazil—for example, the State Forum of AIDS alone included 600 different health/AIDS NGOs (Matthews 2011)—but it had not yet at the time turned its attention to IP. Likewise, negotiations over the Free Trade Area of the Americas (FTAA) and the 1996 WTO Embraer case—a symbol of Brazil's identity as an emerging economic power—had also triggered the mobilisation of civil society groups (Shaffer et al. 2008). Due to the much-publicised WTO pharmaceutical patent case, the merging of these two streams culminated with the creation of a new umbrella civil society organisation—REBRIP (the Brazilian Network for the Integration of Peoples), which includes a number of NGOs, including ABIA,[17] Conectas, Fenafar, GAPA[18] and Gruppo Pela Vida—which would become one of the key actors in the subsequent IP contests (Flynn 2013). Despite such strengthening of civil society groups, however, the initiative for the most important changes to the patent law came, again, from within the state.

Symbolically, the tensions created by the IP orientation of the Brazilian state became visible in the passing of the so-called Sarney's law (Lei no. 9.319) within a few months of the new IP law. Essentially, it guaranteed all AIDS patients free therapeutic care, while the state would pick up the bill. This was not a radical law, for the *movimento sanitário* had already achieved the most radical institutional rupture in Brazil's social policy in 1988 by securing the constitutional right of free and universal health care for all for the first time in the country's history (Fleury 2014). Given the especially active mobilisation of the AIDS movement in keeping the state on its toes in this respect, Sarney deserved little praise—except, perhaps, in persuading an unsupportive FHC not to veto the bill—but Sarney's law did strengthen the status of Brazil's National AIDS Programme, whose success and international reputation the state would skilfully use in the WTO pharmaceutical patent case in 2000 (Nunn 2008; Shaffer et al. 2008; Flynn 2013). Without apparently recognising the tensions inherent in building a national universal healthcare system and following a strategy of handing out monopolies to foreign pharmaceutical companies that controlled the domestic market, FHC's Health Minister boasted: '[T]oday our economy is more open and unprotected than the American one. We did not hesitate to abolish all taxes for the import of medication. We did this to … increase competition' (Serra 2003, quoted in Biehl 2004: 112). All indicators pointed to the opposite: the already high import levels for APIs doubled and those for drugs increased sixfold during the 1990s, drug prices rose 54% above inflation between 1989 and 1999, 81% of government's AIDS drug expenditure was claimed by foreign pharmaceutical companies in 1999 and the trade deficit in pharmaceuticals rose threefold in the space of two years, reaching US$1.277 billion in 1997 (Galvão 2000; Biehl 2004; Mazzoleni and Póvoa 2009; Flynn 2013).

The executive has not, to date, provided the necessary financial base on which to build the universal healthcare system (*Sistema Único de Saúde*, SUS) as stipulated in the Constitution; indeed, total public healthcare expenditure fell from 16.5% to 15.2% of the total social expenditure during 1990–2005 and, despite rising to 3.9% of GDP in 2012, still remains less than half the average in countries with a similar commitment to universal health care (Gadelha et al. 2013: 1616; Leubolt 2014: 9). Insufficient investment and rising healthcare deficits threw in sharp relief both the weak domestic productive base to support the SUS and the negative effects of the overzealous pharmaceutical patent protection offered in the 1996 IP law. In order to reduce the sectoral trade deficit, two Presidential Decrees[19] clarified the conditions and simplified the procedures for issuing compulsory licenses in the public interest—only one would be granted, in 2007—and a Provisional Measure[20] inserted Bolar provisions (Gosain 2007; Portilho and Gosain 2007; Shadlen 2017).[21] The latter was a means of speeding up the entry of generics in the market, and the former a tool of boosting the state's position when negotiating drug price discounts for the SUS. Threats of using compulsory licensing were used regularly and successfully in reducing antiretrovirals prices from 2000 onwards, even though they often remained higher than the lowest international company price. But perhaps the most important change that would become the lifeline of Brazil's national pharmaceutical sector was the creation of the National Health Surveillance Agency (Anvisa), again through a Presidential Decree (no. 3.029/1999). Notably, no remedies could be found for 'pipeline patents'. Evidence that no less than 80% of the National AIDS Programme budget went into purchasing only four drugs benefitting from 'pipeline patents' led to a legal challenge by civil society groups on the patents' unconstitutionality in 2007 (Reis et al. 2009), but the Supreme Court has so far not turned its attention to it.

In spite of being a regulatory agency that lacked an industrial and technological orientation, the creation of Anvisa would simultaneously help improve the fortunes of the domestic pharmaceutical sector during the 2000s and bring to the fore the tensions between Brazil's IP orientation and its ambition of becoming a technological power. These tensions became more prominent with the election of Lula, whose state project would differ from FHC's 'new state' project mainly on account of the more prominent role granted to innovation and technologically intensive industrialisation in making Brazil a knowledge power in the twenty-first century. The inclusion of the pharmaceutical sector as a strategic sector from the first industrial policy (2003) onwards—signalling the moment when the human rights narrative that had been dominant until then would be supplemented by a strong economic rationale—was followed by increased funds to support the public labs and private companies in the sector in the form of dedicated funding streams, primarily from the BNDES and FINEP, which were unprecedented in the history of the sector in Brazil (Shadlen and Fonseca 2013; Shadlen 2017). Aided by other state policies—for example, the (belated) introduction of the generics category by Anvisa in 1999, tax incentives, changes in the government procurement rules and the coordination of relevant policies in the Health–Industrial Complex programme in 2007—the share of generics markets in Brazil grew to 17–18% of the total market at the end of the decade, of which nearly 90% was controlled by Brazilian firms (Gadelha et al. 2013; Caliari and Ruiz 2014).

But all was not good. The deficit in the pharmaceutical sector continued to grow, reaching nearly US$5 billion in 2012, accounting for nearly half of the growing healthcare sectoral deficit, and 48 transnational pharmaceutical companies still accounted for around 80% of the total market by revenues (Flynn 2013; Gadelha et al. 2013). Moreover, some of them (e.g. Roche and Aventis), having had their patent applications rejected, were challenging Anvisa's involvement in the patent grant process in courts as an infringement of TRIPS (Sweet 2013). The same Presidential Decree that had created Anvisa in 1999 had charged it with the task of reviewing all pharmaceutical patents approved by the patent office (INPI) before the patent could be granted by the latter. This so-called 'prior consent' rule and the dual-examination system it set in place from 2001 onwards—clearly with a view to limiting 'evergreening', as the Indian 3(d) section would attempt to do a few years later—became the clearest manifestation of the conflicts between Brazil's IP orientation and its ambition of becoming a technological power during the 2000s. Not only foreign pharmaceutical companies—supported, as earlier, by INTERFARMA and ABPI—but state agencies, too, attacked the 'prior consent' mechanism. The Ministry of Industry and Development and the INPI, seeing incremental innovation and its patenting as the main kind of contribution Brazilian pharmaceutical companies could make to Brazil becoming a competitive knowledge economy, sought to abolish Anvisa's role, even though it had rejected only around 10% of pharmaceutical patents approved by the INPI (Shadlen 2011; Correa et al. 2014). Conflicts intensified in 2009 when Brazil's Attorney General supported a request by the INPI—incidentally, now promoted by the WIPO to the first-ever Latin American International Searching Authority (ISA)—to abolish 'prior consent' and only subsided temporarily when, following a direct intervention from the executive, a precarious 'workflow' between the INPI and Anvisa was arranged in 2013 (Correa et al. 2014; Shadlen 2017), although ambiguities and conflicts have by no means been eliminated.

As conflicts unfolded, civil society groups continued to support Anvisa's 'prior consent', but the same cannot be said about the national pharmaceutical sector (Reis et al. 2009). The lack of support for the 'prior consent' mechanism it had favoured merely a decade ago was notable; indeed, the newly formed *Grupo Farma Brasil* (GFB) has repeatedly voiced opposition to what it considers to be a restrictive pharmaceutical patent regime (Shadlen 2017). Created in 2012, the group represents some of the largest national pharmaceutical companies—for example, EMS Corps, Hypermarcas, Aché, Libbs and Eurofarma, now among the top ten companies in the market—which grew as a result of measures discussed here and whose aim is that of transforming Brazil into an important node in the global pharmaceutical innovation market (GFB 2014).

Although group members have increased their R&D intensity, this still remains low by international standards. Public labs and universities remain the stronghold of pharmaceutical research in Brazil; if patent applications are anything to go by, the rather small and unchanged 5% share of patent applications in the sector by national entities is mainly accounted for by these public bodies.[22] The most notable achievement of GFB members to date has been their expansion outside the Brazilian market from 2009 onwards, predominantly in the Latin American market (Pimentel et al. 2014; Cunha 2015), again helped by state support for the internationalisation of Brazilian companies as part of its 'positive insertion' strategy. Another novelty in this part of the sector has been its emerging involvement with clinical trials (GBF 2014), but this, too, remains small: 77% of the clinical trials approved by Anvisa during 2009–2012 were sponsored by foreign pharmaceutical companies, usually in collaboration with public labs or universities (Silva et al. 2015). Importantly, none of the active pharmaceutical ingredients subject to these clinical trials appeared to be in the national list of key medicines. To make matters worse, as the domestic pharmaceutical market grew—it had already become the world's sixth-largest pharmaceutical market in 2013—so did its appeal to foreign companies, and as a result, a new and ongoing wave of acquisitions has already seen some of the emerging domestic companies bought by foreign pharmaceutical companies (Caliari and Ruiz 2014; Viana et al. 2015).

6.3 Reforming IP Rules for Plant Genetic Resources

Foreign companies' efforts to maintain their position in the domestic pharmaceutical market have been accompanied by efforts to increase the technological rents Brazil's IP law already offers them; for example, they have regularly mobilised to secure second medical use patents, patent linkages and patent extensions for administrative delays, that is, IP forms of the kind they have secured in advanced economies, without success so far (Portilho and Gosain 2007; Reis et al. 2009). Likewise, new bills have been proposed in Congress to limit pharmaceutical patent rights—the most comprehensive of these to date has been *Projeto de Lei* no. 5.402 proposed in 2013 to reconcile patent law with Brazil's competitiveness orientation[23]—again, without success. All is not quiet on Brazil's IP front, for, in addition to pharmaceutical IP, demands to change IP related to biotechnology have also multiplied. In a marked difference to the debate over Brazil's IP law in the early 1990s, some domestic groups—predominantly researchers and representatives of the fledging biotech sector—have couched their demands for changes to IP law in competitiveness terms, although this time in the opposite direction. The two most visible demands have been those for greater access to Brazil's genetic resources and those for stronger biotech IP protection, clashing with the state's control over genetic resources on the former front, but much better aligned with the state's own position on the latter.

These demands have come more visibly to the fore since the biotech sector was labelled a 'gateway to the future' and a 'frontier technology' in the first industrial policy (PITCE 2003); following consultations with the Biotechnology Competitiveness Forum formed that same year, state support for the sector grew and consolidated further with the launch of the Biotechnology Development Policy in 2007, as mentioned earlier. These events signal the point in time when state support for the biotech sector became more systematic and better articulated in its project of transforming Brazil into a technological power, for state's involvement in the sector has a much longer history. The strengthening of the collaboration between the Brazilian state and the Rockefeller Foundation from the 1950s onwards, for instance, laid the ground for two of the most successful applications of traditional biology in Brazil—population genetics and agricultural research—largely concentrated in public universities and public labs, which in the area of agricultural research were reorganised with the creation of Embrapa,[24] the Brazilian Agricultural Research Corporation, in 1973 (Souza and Santos 2014; Nehring 2016). Well before the strengths of these sectors would come to shape the ambition of the state to transform Brazil into a modern biotech powerhouse at the beginning of the twenty-first century (e.g. Furlan et al. 2006; Lula 2007), various programmes and funding streams had been set in place, among them the 1975 *ProÁlcool* Programme, which would eventually contribute to Brazil becoming a key player in the world biofuels market (Nastari 1983). Following the hyperbolic expectations accompanying the emergence of the modern biotech sector in frontier economies, the Brazilian state responded by including biotechnology as one of the key sectors in its Programme for Scientific and Technological Development (PADCT I and PADCT II) during the difficult 1985–1995 period, followed by the advent of the sectoral fund for biotech in 1999 and other more systematic measures from the early 2000s onwards (Schwartzman 1995; Koeller and Cassiolato 2009).

Perhaps the most perceptible sign of the state's commitment to the development of biotechnology can be found in the partial shift of the *Amazônia* imaginary away from the national security, extractivist–developmentalist nexus of the past towards one of *Amazônia* as a treasure trove of biodiversity that would improve Brazil's fortunes in the twenty-first century. Deeper forays into the Amazon continue, of course, driven as much by the state's own extractivist–developmentalist drive as by private actors' illicit acts, and its status as a national security concern and a symbol of national identity has by no means weakened. The point is rather that Brazil's economic crisis and the democratisation of its political and social structures during the 1980s, combined with the concomitant rise of international environmental concerns and hopes over the potential of modern biotechnology, generated a more complex layering of perceptions about the Amazon in the national imagination (Tulchin and Golding 2002; Garfield 2013). If earlier international concerns about deforestation would invariably be framed as attacks on Brazil's sovereignty, a display of *cobiça internacional* (foreign covetousness) over Brazil's resources and a ploy to take control over the Amazon, in the late 1980s, they started being framed in a different light; the most spectacular sign of this change was Collor playing host to the Earth Summit in 1992, from where the CBD emerged, so as to showcase Brazil's commitment to environmental protection (Keck 2002; Garfield 2013). Likewise, the traditional and indigenous communities living in the Amazon, often seen as a hindrance to the country's development and, when aligned with international human rights and environmental NGOs, as dupes of foreigners seeking to take control of Amazonian riches, now emerged as defenders of a national biogenetic patrimony of untold economic value[25] and as a fountain of knowledge that could propel Brazil into knowledge economy status (Cunha and Almeida 2000; Conklin 2002). A number of programmes and agencies to transform Amazon's biodiversity into innovative products were created; this activity was not novel, for by the mid-1980s, a significant system was already in place, largely due to the lobbying efforts of a small group of Brazilian scientists and conservationists, who had convinced consecutive governments of the potential importance of Amazon's resources for the biotechnology and pharmaceutical industries (Garfield 2013). But agencies and programmes on Amazon's biodiversity proliferated from the 1990s onwards, for example, the launch of the Network for the Conservation and Use of Genetic Resources of Amazônia in the early 1990s (in line with the CBD's approach of 'conservation through utilisation'), the Brazilian Molecular

Ecology Program for the Sustainable Use of Amazonian Biodiversity (PROBEM) in the late 1990s—perhaps the most significant initiative to convert Amazon's biodiversity into a source of high-value-added products and advanced scientific knowledge—and, related to it, the creation of the Biotechnology Centre of Amazônia in 2002 (Coutinho et al. 2001; Gouvea and Kassicieh 2005; Furlan et al. 2006; Bound 2008).

Following Brazil's ratification of the CBD in 1994—the first country to do so—a number of projects were initiated primarily in the Amazon basin that generated hardly any economic benefits but plenty of headaches for the FHC administration. One particularly problematic front emerged in the form of indigenous communities' resistance against such projects taking place without their consent, not to mention the fact that many (but not all) still consider the idea of genetic resources' commercialisation as incompatible with their traditions (Eimer et al. 2016). A second front of resistance reinforcing the first emerged in the form of law proposals on the CBD's implementation that sought to make its optional endorsement of indigenous groups' consent into a mandatory requirement. The most influential of these was made in 1995 by Congresswoman Marina Silva—a notable environmental figure in Brazil—followed by even further-reaching proposals by PT representatives (Velez 2010). Such proposals clashed with FHC's aim of preserving executive's powers over this matter, and although all state agencies agreed that genetic resources should be utilised towards economic goals, the degree of indigenous communities' involvement was strongly disputed and supported primarily by the Ministry of the Environment and the public persecution office (Eimer et al. 2016). It is difficult to evaluate how long this turf war would have lasted; due to widespread anger about the plundering of Brazil's biodiversity that erupted on evidence of an unfavourable contract signed between BioAmazônia—a social organisation that had been created to implement PROBEM—and Novartis, FHC was forced to issue a Provisional Measure (no. 2.052) in 2000 which effectively implemented the CBD and, through numerous resolutions, ushered in an access and benefit-sharing system in Brazil[26] (Coutinho et al. 2001).

The contradictions inherent in the state's aim of simultaneously exerting its control over Brazil's genetic resources and supporting the development of its biotech sector emerged, first, in the prohibition of patent protection for naturally occurring life forms and, second, in a rather restrictive access and benefit-sharing regime ushered in by FHC's Provisional Measure. Following the latter, an inter-ministerial structure, the Council for Managing Genetic Patrimony (CGEN), took control over access to genetic resources in 2002, marking the point in time when various groups, key among them Brazilian scientists and researchers, started criticising the 'heavy-handed' approach as contributing to Brazil's loss of competitiveness (BioMinas 2011; Filoche 2012). Initially, the object of criticism was the near-impossibility of accessing genetic resources, but this argument lost its strength as the state progressively and silently loosened its grip and Brazilian researchers secured preferential access to them. Soon after the CGEN became operative, the Brazilian Institute for the Environment and Renewable Natural Resources (IBAMA) became responsible for considering applications to access genetic resources for non-commercial purposes and issued not only a large number of authorisations but also 'special dispensations' for Brazilian public research institutes, such as Embrapa, the Butantan Institute and others (Filoche 2012). Other state agencies got progressively involved in issuing authorisations, totalling around 1300 in 2013, of which nearly half were for research purposes (Gross 2014: 23). In light of this shift, criticism moved to the manner in which Brazil's IP system forbade Brazilian researchers and companies to patent inventions based on the nation's biodiversity, whereas foreign companies (whose access remains conditional upon formal cooperation with a national entity) could do so in countries where it was possible, for example, the US (Octaviani 2010; Filoche 2012). In such framing, widely supported by the nascent (private) biotech sector and the ABPI, concerns about loss of competitiveness on the part of Brazilian entities gained more strength by virtue of appearing simultaneously as a fight against biopiracy. Few noted that Brazilian companies themselves, encouraged by state's policies aimed at improving Brazil's rating in international patenting tables, had no hesitation in patenting inventions based on the nation's biodiversity abroad (Chamas et al. 2007; Filoche 2012).

Following the continuous and more systematic state support for the sector since 2003, the number of private biotechnology companies has grown in Brazil, predominantly in areas of human health and agriculture (Biominas 2011). Despite (or, likely, because of) such growth, the most dynamic and research-intensive companies get promptly bought out by foreign companies—as was the fate of the first and rather successful Brazilian biotech company, Biobrás, established as a spin-off of the Universidade Federal de Minas Gerais in 1971 and bought by Novo Nordisk in 2001—leaving in their wake diminished local innovation systems and companies that tend to perform routine and low-tech activities (Cassiolato et al. 2011). As in the case of the pharmaceutical sector, the biotech sector is largely controlled by foreign companies, who account for over 90% of biotech patent applications in Brazil.[27] Universities and public research institutes account for over half of the relatively small pool of resident patent applications and state agencies are here, as in the pharmaceutical sector, the most dominant knowledge market players.

Unlike most frontier economies and ambitious developing countries (e.g. India), Brazil's 1996 IP law does not allow patents on naturally occurring life forms such as gene sequences and microorganisms, unless modified by human intervention. This represented a reversal of fortunes, for Collor's original 1991 draft had generally been less obliging to those agitating for strong IP protection standards, with the exception of biotechnology: draft provisions allowed patents on a wide range of life forms on account of arguments presented by the ABRAPI and others that Brazil's well-developed agricultural biotechnology would enable it to make the leap into 'modern' biotechnology markets (Shadlen 2017). Nevertheless, this argument appeared less convincing when Embrapa, *the* leading agricultural research institution, entered the fray on the side of the FLUC (Forum for the Free Use of Knowledge) and other opponents of the bill, arguing that such a broad scope would likely inhibit its researchers' access to new biotechnologies (Filomeno 2014b; Shadlen 2017).

Contests over 'life patents' during this period dominated those over pharmaceutical patents and brought together in opposition, as discussed earlier, a wide coalition of groups that included environmental NGOs, trade unions, scientific societies and institutes, and the Catholic Bishops Conference (Newell 2008; Filomeno 2014b; Shadlen 2017). The restriction on 'life patenting' in the 1996 IP law does not appear to have negatively affected the dynamism of Brazil's biotech sector; on the contrary, perhaps the most notable success story that put Brazil's life science sector on the global map of bacterial and plant pathogens genomics, ONSA,[28] was based on a collaborative structure, in this case opting to emulate the EU collaborative model, as opposed to the hierarchical one that characterises the US biotech sector (Harvey and McMeekin 2005). With ONSA achieving a worldwide reputation, this collaborative model was adopted elsewhere in the sector not without success. Recent demands to expand patentable matter in the name of competitiveness—incidentally, most forcefully made by successful Brazilian biotech firms such as Allelyx and CanaVialis[29] that emerged out of the ONSA network in the early 2000s—come at a critical juncture because, if they succeed, this characteristic of the sector may be eliminated. They may indeed succeed for the state, for its part, also supports the view that greater IP protection is key to the further development of the biotech sector. The 2007 Biotechnology Development Policy made clear, for instance, that it was fundamental to spread the IP culture and mobilise individuals' skill and institutional mechanisms towards the effective use of IP so as to improve the competitiveness of Brazil's biotech sector (Octaviani 2010).

This position signals a shift compared with the mid-1990s, when concerns that the national biotech sector was not strong enough to compete with foreign companies, which would, in turn, appropriate the national genetic patrimony through patenting, was also a key factor in the 1996 IP law prohibiting patent protection for naturally occurring life forms (Chamas et al. 2007; Filoche 2012). Similar concerns appear to have guided the formulation of a seemingly obscure law—the Biosafety Law of 1995—that would become the eye of the storm over GM crops in Brazil during the late 1990s. As in India, contests over GM escalated due to Monsanto's activities, in the case of Brazil following Monsanto gaining approval to commercialise its GM Roundup Ready (RR) soy in 1998, bolstered in no small measure by its RR technology patent acquired thanks to the 'pipeline' mechanism (Filomeno 2014a). Immediately followed by a lawsuit initiated by a group of anti-GM NGOs—among them the Institute of Consumer Defense (IDEC) and Greenpeace—this event catalysed a trenchant battle over GM that would sweep across state agencies, the judiciary, civil society groups and farmers in their fields (Scoones 2005; Newell 2008; Peschard 2012). Despite the anti-GM rhetoric while in opposition, the five-year legal moratorium on the commercial production and release of GM crops was eventually overturned by a Presidential Decree issued in the first year of the Lula administration to allow the (illegal) GM harvest of the year, followed by a new biosafety law draft sent to Congress with the aim of centralising control of the GM policy in the hands of the executive (Jepson 2002). The new biosafety law draft became the key battlefront not only on GM issues, but also on those related to stem cell research, inserted in the draft at the behest of the respective scientific community. Such strategic coupling had the effect of framing arguments from the unusual alliance between those favouring a more cautious GM approach and the strong *bancada evangélica*[30] against stem cell research as backward, hindering the adoption of solutions to better health, eradication of hunger, pest control and other issues (Cesarino and Luna 2011; Reis-Castro and Henrickx 2013).

The new biosafety law passed in March 2005, decisively turning the tide against the anti-GM campaign and burying its vision of a free-GM Brazil. This movement, although strong, had since the start been constrained in a context where arguments about GM and agro-biotech were framed either in technical terms (e.g. biosafety) or in economic competitiveness terms; thus, it never fully succeeded in shifting attention to issues related to unequal relations of power, land ownership or property rights (Scoones 2005; Newell 2008). The 1995 biosafety law had intentionally limited GM-related issues to technical ones, at once making it harder for dissenting voices to gain ground and appeasing the concerns of those the FHC administration saw as the most important constituencies in the debate: the agribusiness and scientific communities. The interests of the latter were safeguarded by provisions that permitted and regulated transgenic research—certainly with an eye on Embrapa—as a means of supporting Brazilian scientific competitiveness, while those of the former by allowing commercial products that contained GM traits, but prohibiting the planting of GM crops, with a view to allowing time for Brazilian agro-biotech firms to develop their own seed varieties and competitive strengths (Jepson 2002). That the Brazilian executive supported transgenics during this period is visible not only in the relatively large number of GM experiments approved, but also in the resolve to silence conflicting positions on the question of agro-biotech among ministries. One such instance was the issuing of a statement clarifying government's support for biotech, signed in 2000 by the previously warring ministers of Health, Agriculture, Justice, Environment and S&T (Jepson et al. 2008). Whatever hopes the Free-GM Brazil movement had pegged on the new Lula administration, they were dashed as soon as Roberto Rodrigues, a well-known agribusinessman, was appointed as the new minister for agriculture in 2003. Since the new biosafety law of 2005, over 28 GM varieties of soybean, cotton and corn have been approved, making Brazil the second-largest producer of GM soy worldwide by 2011 and the developing country with the largest GM crop area in the world by 2015 (Peschard 2012; James 2015).

As GM seeds are mainly suitable for highly capitalised agro-industrial farming, such transformation could not have happened without the state first creating the conditions for the expansion of the agricultural frontier into the *cerrado*, Brazil's vast savannah interior, which opened up more than 100 million hectares, reaching deep into the northeast and the Amazon regions

(Wilkinson and Herrera 2010). Often depicted as an achievement on the part of the Embrapa in adapting soy and other crops to this particular environment, the transformation of the *cerrado* from an area whose acidic soils were considered unfit for modern agriculture into one of the most important conquests of modern agriculture of the last century was achieved in significant measure by decades-long involvement of Rockefeller scientists, US technological packages and Brazilian state-sponsored infrastructure projects, generous credit lines and state-supported migration of industrial farmers from the south (Nehring 2016). From the very beginning, no alternatives to this agro-industrial colonisation were ever entertained and its design was both aimed at the international market and dependent on it for inputs to maintain productivity (Rada 2013). Unsurprisingly, the *cerrado* is the most input-dependent agricultural zone in Brazil, accounting for nearly half the national fertiliser and pesticide expenditures; over 70% of the pesticide market is controlled by eight agrochemical corporations (including Monsanto, Syngenta, Bayer, BASF and DuPont) and nearly 70% of national fertilisers—a large part of which is claimed by *cerrado*'s key export crops of soy and corn—are imported from abroad (Nerhing 2016: 15). Nor did the expansion of this agricultural frontier help solve the perennial issue of land concentration. Recognised as a problem that required an expeditious solution as early as the 1946 Constitution, redistributive land reform has been part of all governments' programmes, all of which have more or less failed: under 1% of farms produce over 50% of gross agricultural income and occupy more than 43% of the total area, while farms of less than 10 hectares, deemed insufficient to generate income, account for nearly half the number of farms but for only around 3% of the total area, not to mention nearly 3 million families classed as landless (Sauer and Leite 2012: 876; Mueller and Mueller 2014: 15).

The tensions emanating from the historical neglect of the small-scale and family farming sector and support for the agribusiness sector that bore fruit in its conquering of world markets manifested themselves in contests over the new seed and PBRs legislation contemplated during the 1990s. Until the early 1990s, Embrapa and other public research institutes had opposed PBRs on grounds that agricultural R&D ought to remain in the public domain and out of fear that their introduction would facilitate the takeover of the nascent domestic sector by foreign companies (Filomeno 2014a). Following the new state orientation, Embrapa reformulated its mission as a state-owned corporation in 1996 to include the imperative of obtaining profits from the exploitation of its research outcomes, a change that was instrumental in the passing of the 1997 Law of Protection of Cultivars (Lei no. 9456) (Filomeno 2014a; Peschard 2012). Having lost the support of the largest holder of cultivars and the largest agricultural research institute in the country, resistance against the bill on the part of CONTAG (the National Confederation of Rural Workers) and the MST (the Rural Landless Workers movement) weakened; a provision on farmers' right to save as well as exchange and sell seeds to other small farmers was added by the Senate but promptly removed by the Chamber, although the final text, being largely based on UPOV 1978, did incorporate the right of rural producers to save seeds (Filomeno 2013, 2014b). The cultivar law of 1997 had the effect of increasing the appeal of the Brazilian market to foreign seed companies; Monsanto, for instance, entered the market that same year through purchasing FT Sementes and Agroceres, the main Brazilian seed companies at the time, and establishing a partnership with Embrapa itself.

Despite increased participation in the seed market by transnational companies, public research institutions had more or less the same share of protected varieties in the market as foreign companies (around 39%) by the mid-2000s. Embrapa remains without doubt the main market player: it alone owned 27% of protected cultivars and nearly 41% of the total when its partnerships were included (Chamas et al. 2007). Correspondingly, its revenues from royalties increased rapidly each year; by 2011, they were around 1.14% of its budget, as compared with an average of 0.9% for US public research institutions (Filomeno 2014a: 85). Having become the major player in the seed and agro-biotech knowledge market, this public institution would play a key role in limiting whatever farmers' rights the 1997 cultivar law had provided. What had started as an effort to regulate the production and trade of seeds, for instance, ended up at Embrapa's behest in restricting the right to save seeds in the 2003 Law of Seeds and Seedlings and, more broadly, in limiting the role that the informal seed system—still serving the majority of farmers in Brazil—and farmers' own seed selection and innovation plays in agro-biodiversity (Santilli 2012; Filomeno 2014a). In 2002 and 2009, the Ministry of Agriculture, supported by Embrapa, also sought to amend the 1997 cultivar law in order to make it more generous to IP-holders using arguments about the need to encourage R&D investment, fight against seed piracy and help make Brazil a leader in agro-biotech (Filomeno 2014a). No less than three bills, one originating from the executive, were under review in 2012, once again aiming to expand the rights afforded to IP-holders by the 1997 cultivar law and restricting seed-saving practices (Peschard 2012).

Resistance against these moves has recently been bolstered by some large soy agribusinesses turning against Monsanto on account of its RR 'technological fee',[31] accompanied by calls that the state increase its investment in agricultural R&D as a matter of national security and sovereignty to counteract the control exercised by private transnational companies in the seed market (Peschard 2012; Filomeno 2013). For its part, the state has not intervened in farmers' disputes with Monsanto, and as far as agricultural R&D funding is concerned, it had been doing just that. Driven less by concerns of national security than with turning Brazil into a leader in agro-biotech, Embrapa's budget accounted for two-thirds of the entire Latin American public agricultural research budget by the late 1990s, and by 2010, it had enjoyed a real increase of 70% in its budget over a decade (Jepson 2002: 912; Filomeno 2014a: 103). As its control over cultivars and technology in the Brazilian and Latin American markets grew, Embrapa became the undisputable crown jewel of this strategy. Other clear beneficiaries have been large Brazilian agricultural companies, whose fortunes improved during the 1990s and 2000s.

It is widely accepted today that the potential opened up by biotech can liberate agriculture from its 'low-tech' categorization. Despite some successes of Brazilian agriculture in this regard, the Brazilian state remains keenly aware of its reputation as primarily a commodity exporter. Shortly before the last PT administration was toppled in August 2016, its Minister of S&T had an opportunity to acknowledge this fact and reinstate the strategy laid out by the first PT administration 13 years ago: '[T]oday, everyone says that Brazil is a commodity exporter. We'd like to stop exporting raw material and become a country that exports innovation' (Pansera 2016). Disappointment, mixed with hope for the future, characterises the record not just of the agricultural sector, but of the broader goal of transforming Brazil into a knowledge power. As the new decade started and notwithstanding GDP growth rates of around 4.5% per annum, it was clear that despite the state's efforts to guide Brazil's positive insertion in the global economy through strengthening its structural competitiveness—a growth strategy, as discussed, laid out as early as the mid-1980s—accumulation had decisively shifted to the financial sector and natural resource extraction/production. The ratio of financial assets to productive assets reached nearly 75%, up from 15% in 1992; the manufacturing share of GDP was 14.5%, a level similar to that in 1956; and the share of high-tech manufacturing still represented only 7% of Brazil's total merchandise exports (Paulani 2010: 369; Palma 2012). Becoming the technological and environmental power of the twenty-first century remains a project for Brazil.

Notes

1. A recent instance of this can be found in 'The Amazon is ours. It is not for sale ...', a letter co-authored by the Brazilian Foreign Minister, Environment Minister and the Science and Technology Minister, published by the *Independent* on 31 October 2006.
2. Machado de Assis; see Daniel (2014).
3. BNDES is Brazil's National Economic and Social Development Bank.
4. IEDI is the Institute of Studies for Industrial Development, a think tank based in São Paulo, with strong links to the main national manufacturing firms.
5. CEPAL or ECLA, the UN Economic Commission for Latin America and the Caribbean.
6. INPI is Instituto Nacional da Propriedade Industrial.
7. Sixteen funds were launched in 1999, including areas of biotech, agribusiness, pharmaceuticals and *Amazônia*.
8. FINEP, the Funding Authority for Studies and Projects, was established in 1967 with the goal of establishing and financing modernisation and industrialisation projects.
9. PITCE is the Industrial, Technological and Foreign Trade Policy.
10. INTERFARMA (Associação da Indústria Farmacêutica de Pesquisa) is the Assocation of the Research Pharmaceutical Industry in Brazil.
11. ABRAPI (Associação Brasileira de Biotecnologia) was the Brazilian Association of Biotechnology, replaced in 2014 by the ABBI (Associação Brasileira de Biotecnologia Industrial).
12. CNI (Confederação Nacional Da Indústria) is the National Confederation of Industry in Brazil.
13. ABIFINA (A Associação Brasileira da Indústria de Química Fina, Biotecnologia e suas Especialidades) is the Brazilian Association of Fine Chemical and Biotechnological Industries (and related sectors), founded in 1986.
14. ALANAC (Associação dos Laboratórios Farmacêuticos Nacionais) is the Association of National Pharmaceutical Laboratories, founded in 1983.
15. Bolar provisions are research exemptions to patent rights especially relevant to drugs; essentially, they permit research and testing for regulatory approval for a generic drug before a patent expires without constituting an infringement of said patent.
16. Pipeline patents offered full protection for pharmaceuticals that were patented (but still awaiting marketing approval) elsewhere *before* the entry into force of the TRIPS Agreement (1995). If a pharmaceutical product is patented in 1993 in the US (but is still waiting marketing approval), then it would not have benefitted from patent protection as stipulated in TRIPS in countries not required to comply with TRIPS yet, say, in Brazil, when patents had to be provided for pharmaceutical products only in 2005.

17. ABIA (Observatório Nacional de Políticas de AIDS) is the National Observatory of AIDS Policies in Brazial.
18. GAPA (Grupos de Apoio à Prevenção à Aids) is an umbrella organisation of NGOs that support the prevention of AIDS.
19. No. 3.201/1999 and no. 4.830/2003. Presidential decrees do not, as a matter of course, constitute law reforms, but sometimes, they can introduce substantial changes.
20. No. 2014-1 issued in 1999. Provisional measures, also issued by the president in 'relevant and urgent cases', have the force of law, but must be submitted to Congress for approval usually within 60 days, lest they expire.
21. All these measures and changes were incorporated into the IP law in 2001.
22. Data from the INPI Badepi version 1.1 database, collected and analysed by Dr. Leonardo Costa Ribeiro of INMETRO (Instituto Nacional de Metrologia, Qualidade e Tecnologia), in file with author.
23. *A Revisão da Lei de Patentes: Inovação Em Prol da Competitividade Nacional* (Patent Law Reform: Innovation Towards National Competitiveness). The proposal is available here: http://bd.camara.gov.br/bd/handle/bdcamara/14796, last accessed 20 September 2016.
24. Embrapa, the Brazilian Corporation for Farming and Livestock Research, was created in 1973 to organise and expand public research on agriculture that until then had been decentralised.
25. Despite the impossibility of calculating such value, the Institute of Applied Economic Research (IPEA) tallied it to US$ 2 trillion (Izique 2002).
26. Despite Brazil being the first to ratify the CBD in 1994 and playing a key role in the negotiation of the Nagoya Protocol during the first decade of the noughties, many in Brazil were unsupportive of ratifying because the national access and benefit-sharing system that was put in place after the 2000 Provisional Measure was deemed by many as cumbersome and ineffective. Attempts to create a new biodiversity law on the part of the executive branch since 2009, however, were repeatedly resisted by agribusiness interest groups, supported by the influential *bancada ruralista* in Congress and the Ministry of Agriculture in government, largely on the grounds that the proposed rules—and the Nagoya Protocol—may lead to claims of benefit-sharing for genetic resources on which Brazil's agribusiness success is built, the vast majority of which (e.g. soy, sugar, microorganisms used in the food and biofuel sectors etc.) are not native to Brazil. Although the new biodiversity law that finally passed in 2015 swung the pendulum unmistakably in favour of those in the business of manipulating genetic resources as compared with the earlier Provisional Measure, it did not alleviate agribusiness concerns, while exacerbating those of indigenous communities, which continue to resist its implementation (Welch 2015).
27. Data refers to the post-1996 period; see endnote 13.
28. ONSA, the Organisation of Nucleotide Sequencing and Analysis, was established in the São Paulo state in 1997 to map the yeast genome.

29. Alellyx was formed in 2002 and CanaVialis in March 2003; both were bought by Monsanto in 2008.
30. Bancada evangélica refers to a growing group in Congress composed of congressmen of different political parties united in their evangelical faith.
31. This refers to a fee Monsanto was collecting from farmers using its RR soy seeds in Brazil due to its RR technology patent; this was distinct from the royalties collected from the use of RR seeds, thus constituting a dual-payment regime, which has been the focus of legal challenges in Brazil since the mid-2000s.

Bibliography

Ackerman, F. 1971. Industry and Imperialism in Brazil. *Review of Radical Political Economics* 3: 1–39.

Biehl, J. 2004. The Activist State: Global Pharmaceuticals, AIDS and Citizenship in Brazil. *Social Text* 22 (3): 105–132.

BioMinas. 2011. *The Brazilian Life Science Industry: Pathways for Growth*. Report available at http://biominas.org.br/en/downloads/the-brazilian-life-science-industry-pathways-for-growth-biominas-e-pwc/. Accessed 20 September 2016.

Bound, K. 2008. *Brazil: The Natural Knowledge Economy*. London: Demos.

Caliari, T., and R.M. Ruiz. 2014. Brazilian Pharmaceutical Industry and Generic Drugs Policy: Impacts on Structure and Innovation and Recent Developments. *Science and Public Policy* 41 (2): 245–256.

Carvalho, J.M. 2000. Dreams Come Untrue. *Daedalus* 129 (2): 57–82.

Cassiolato, J.E., and H. Schmitz, eds. 1992. *Hi-Tech for Industrial Development: Lessons from the Brazilian Experience in Electronics and Automation*. London: Routledge.

Cassiolato, J.E., G.F. Zucoloto, M.S. Rapini, and S.G.A. Souza. 2011. The Recent Evolution of the Biotech Local Innovation System of Minas Gerais: University, Local Firms and Transnational Corporations. In *Biotechnology and Innovations Systems: The Role of Public Policy*, ed. B. Göransson and C.M. Pålsson, 13–57. Cheltenham: Edward Elgar.

Cassiolato, J.E., G. Zucoloto, R. Molagres, and F. Stalliveiri. 2014. Transnational Corporations and the Brazilian National System of Innovation. In *Transnational Corporations and Local Innovation*, ed. J.E. Cassiolato, G. Zucoloto, D. Abrol, and L. Xielin, 68–132. Abingdon: Routledge.

Cederlöf, G., and K. Sivaramakrishnan. 2006. *Ecological Nationalisms: Nature, Livelihoods, and Identities in South Asia*. Seattle, WA: University of Washington Press.

Cesarino, L., and N. Luna. 2011. The Embryo Research Debate in Brazil: From the National Congress to the Federal Supreme Court. *Social Studies of Science* 41 (2): 227–250.

Chamas, C.I., S. Paulino de Carvalho, and S. Salles-Filho. 2007. Current Issues of IP Management in Health and Agriculture in Brazil. In *Intellectual Property Management in Health and Agricultural Innovation: A Handbook of Best Practices*, ed. A. Krattiger, R.T. Mahoney, L. Nelsen, et al., 1563–1575. Oxford: MIHR.

Conklin, B.A. 2002. Shamans Versus Pirates in the Amazonian Treasure Chest. *American Anthropologist* 104 (4): 1050–1061.

Correa, M., P. Villardi, and P. Villela 2014. Pharmaceuticals and Public Health in Brazil: Copying Essential Drugs, Knowledge Acquisition and Innovation Projects in Public-Private Industrial Networks. Unpublished draft, in file with author.

Coutinho, L.G. 2003. Macroeconomic Regimes and Business Strategies: An Alternative Industrial Policy for Brazil in the Wake of the 21st Century. In *System of Innovation and Development: Evidence from Brazil*, ed. M.L. Maciel. Cheltenham: Edward Elgar.

Coutinho, M., R.L.M. Silva, and M.A. Marin. 2001. Biodiversity and Public Policy Issues in Development: Brazil as a Case Study. Paper Presented at the International Society for History, Philosophy and Social Studies of Biology, Annual Meeting, Hamden, USA.

Cunha, J. 2015. Farmacêuticas Brasileiras Compram Empresas No Exterior por Inovação. *Folha de S.Paolo.* Available at http://www1.folha.uol.com.br/mercado/2015/06/1644256-farmaceuticas-brasileiras-compram-empresas-no-exterior.shtml. Accessed 20 September 2016.

Cunha, M.C., and M. Almeida. 2000. "Indigenous People, Traditional People and Conservation in the Amazon," in Brazil: Burden of the Past, Promise of the Future. *Journal of the American Academy of Sciences* 129 (2): 315–338.

Daniel, G.R. 2014. Machado de Assis and Brazilian Literary Independence: Toward a Postcolonial National Aesthetic. *Machado de Assis Em Linha* 7 (14): 163–182.

Das, A.N. 1994. Brazil and India: Comparing Notes on Liberalisation. *Economic and Political Weekly* 29 (8): 416–419.

Drahos, Peter. 2010. *The Global Governance of Knowledge*. Cambridge: Cambridge University.

Drèze, J., and A. Sen. 1989. *Hunger and Public Action*. Oxford: Oxford University.

Eimer, T.R., S. Lütz, and V. Verena Schüren. 2016. Varieties of Localization: International Norms and the Commodification of Knowledge in India and Brazil. *Review of International Political Economy* 23 (3): 450–479.

Evans, P. 1979. *Dependent Development: the Alliance of Multinational, State and Local Capital in Brazil*. Princeton, NJ: Princeton University Press.

Feijo, C.A., and M.T. Lamonica. 2010. The Importance of the Manufacturing Sector for Brazilian Economic Development. *CEPAL Review* 102: 7–26.

FHC (Fernando Henrique Cardoso). 2000. Discurso do presidente da República na cerimônia de apresentação do "Avança Brasil": Plano Plurianual 2000/2003 e do Orçamento da União para o ano 2000, Palácio do Planalto, Brasília. Available at http://www1.folha.uol.com.br/fsp/brasil/fc0109991l.htm. Accessed 22 September 2016.

———. 2002. O Brasil a caminho da sociedade do conhecimento. In *O Brasil e a Economia do Conhecimento*, ed. J.P. Velloso. Rio de Janeiro: J. Olympio.

Filoche, G. 2012. Biodiversity Fetishism and Biotechnology Promises in Brazil: From Policy Contradictions to Legal Adjustments. *The Journal of World Intellectual Property* 15 (2): 133–154.

Filomeno, F. 2013. State Capacity and IP Regimes: Lessons from South American Soybean Agriculture. *Technology in Society* 35: 139–152.

———. 2014a. *Monsanto and Intellectual Property in South America*. Basingstoke: Palgrave Macmillan.

———. 2014b. Patterns of Rule-Making and Intellectual Property Regimes. *Comparative Politics* 46 (4): 439–458.

Fleury, S. 2014. Building Democracy in an Emerging Society: Challenges of the Welfare State in Brazil. In *Brazil Emerging: Inequality and Emancipation*, ed. J.N. Pieterse and A. Cardoso, 11–31. Abingdon: Routledge.

Flynn, M. 2011. Corporate Power and State Resistance: Brazil's Use of TRIPS Flexibilities. In *Intellectual Property, Pharmaceuticals and Public Health: Access to Drugs in Developing Countries*, ed. K. Shadlen, 149–177. Cheltenham: Edward Elgar Publishing.

———. 2013. Origins and Limitations of State-based Advocacy: Brazil's AIDS Treatment Program and Global Power Dynamics. *Politics and Society* 41 (1): 3–28.

Furlan, L.F., A. Álvares, L.C.G. Pinto, and S. Rezende. 2006. Futuro Certo e Promissor da Biotecnologia. *Folha de São Paulo*, June 23.

Furtado, C. 1992. *Brasil: A Construção Interrompida*. São Paolo: Paz e Terra.

———. 1997. *Obra Autobiográfica de Celso Furtado*. São Paolo: Paz e Terra.

Gadelha, C.A.G., L.S. Costa, J.M.S.V. Maldonado, and M.A. Vargas. 2013. The Healthcare Economic-Industrial Complex: Concepts and General Characteristics. *Health* 5 (10): 1607–1621.

Galvão, J. 2000. *A AIDS No Brasil: A Agenda de Construção de Uma Epidemia*. São Paulo: Editora 34.

Garfield, S. 2013. *In Search of the Amazon: Brazil, the United States and the Nature of a Region*. London: Duke University Press.

Gereffi, G. 1983. *The Pharmaceutical Industry and Dependency in the Third World*. Princeton, NJ: Princeton University Press.

GFB (Grupo Farma Brasil). 2014. *The Rise of a National Biopharmaceutical Industry in Brazil: Consolidation, Trends and Challenges*. Available at https://c.ymcdn.com/sites/casss.site-ym.com/resource/resmgr/CMC_LATAM_Speaker_Slides/2014_CMCLAT_ArcuriReginaldo.pdf. Accessed 21 September 2016.

Gontijo, C. 2005. *Changing the Patent System: From the Paris Convention to the TRIPS Agreement*. Global Issue Paper No. 26, The Heinrich Böll Foundation, Berlin.

Gosain, R. 2007. In Good Health? Compulsory Licenses. *Patent World* 194: 21–30.

Gouvea, R., and S. Kassicieh. 2005. Using Resources in R&D Policy Planning: Brazil, the Amazon and Biotechnology. *Technological Forecasting & Social Change* 72: 535–547.

Gross, R. 2014. *National Study on ABS Implementation in Brazil*. The ABS Capacity Development Initiative and the Government of Brazil. Available at http://www.abs-initiative.info/fileadmin/media/Knowledge_Center/Pulications/ABS_Dialogue_042014/National_study_on_ABS_implementation_in_Brazil_20140716.pdf. Accessed 25 January 2017.

Harvey, M., and A. McMeekin. 2005. Brazilian Genomics and Bioinformatics: Instituting New Innovation Pathways in a Global Context. *Economy and Society* 34 (4): 634–658.

Hurrell, A. 2005. The United States and Brazil: Some Comparative Reflections. In *The United States and Brazil: A Long Road of Unmet Expectations*, ed. M. Hirst, 75–90. New York: Routledge.

Izique, C. 2002. Actions Against Biopiracy. *Revista Pesquisa FAPESP*, June Issue. Available at http://revistapesquisa.fapesp.br/en/2002/06/01/actions-against-biopiracy/. Accessed 22 September 2016.

James, C. 2015. 20th Anniversary (1996 to 2015) of the Global Commercialization of Biotech Crops and Biotech Crop Highlights in 2015. ISAAA Brief No. 51. Available at http://isaaa.org/resources/publications/briefs/51/executive summary/default.asp. Accessed 6 June 2016.

Jepson, W.E. 2002. Globalization and Brazilian Biosafety: The Politics of Scale Over Biotechnology Governance. *Political Geography* 21: 905–925.

Jepson, W.E., C. Brannstrom, and R.S. de Souza. 2008. Brazilian Biotechnology Governance: Consensus and Conflict Over Genetically Modified Crops. In *Food For the Few: Neoliberal Globalism and Biotechnology in Latin America*, ed. G. Otero, 217–242. Austin: University of Texas Press.

Keck, M. 2002. Amazônia in Environmental Politics. In *Environment and Security in the Amazon Basin*, ed. J.S. Tulchin and H.A. Golding, 31–52. Washington, DC: Woodrow Wilson International Center for Scholars.

Koeller, P., and J.E. Cassiolato. 2009. Achievements and Shortcomings of Brazil's Innovation Policies. In *BRICS and Development Alternatives: Innovation Systems and Policies*, ed. J.E. Cassiolato and V. Vitorino, 35–71. London: Anthem Press.

Koeller, P., and J.L. Gordon. 2013. Brazil. In *The State and the National System of Innovation: A Comparative Analysis of the BRICS Economies*, ed. M. Scerri and H.M.M. Lasters, 23–79. London: Routledge.

Kohli, A. 2004. *State-Directed Development: Political Power and Industrialization in the Global Periphery*. Cambridge: Cambridge University Press.

Kolk, A. 1996. *Forests in International Environmental Politics: International Organisations, NGOs, and the Brazilian Amazon*. Utrecht: International Books.

Lamounier, B. 1989. Brazil; Inequality Against Democracy. In *Democracy in Developing Countries*, ed. L. Diamond et al., vol. 4. Boulder, CO: Lynne Rienner Publisher.

Lastres, H.M.M., J.E. Cassiolato, and M.L. Maciel. 2003. Systems of Innovation for Development in the Knowledge Era. In *Systems of Innovation and Development*, ed. J.E. Cassiolato, H.M.M. Lastres, and M.L. Maciel, 1–32. Cheltenham: Edward Elgar Publishing.

Leamon, A., and J. Lerner. 2012. Creating a Venture Ecosystem in Brazil: FINEP's INOVAR Project. Working Paper 12-099, May 8, 2012, Harvard Business School.

Leiva, F.I. 2008. *Latin American Neostructuralism: The Contradictions of Post-Neoliberal Development*. Minneapolis: University of Minnesota Press.

Leubolt, B. 2014. *Social Policies and Redistribution in Brazil*. Working Paper No. 26. Geneva: International Labour Office.

Lula (Luiz Inácio da Silva). 2007. Discurso do Presidente da República, Luiz Inácio Lula da Silva, na cerimônia de lançamento da Política de Biotecnologia, Palácio do Planalto, Brasília, 08 de fevereiro de 2007. Available at http://www.biblioteca.presidencia.gov.br/presidencia/ex-presidentes/luiz-inacio-lula-da-silva/discursos/

Marques, M.B. 1999. Gestão, planejamento e avaliação de políticas de ciência e tecnologia: hora de rever? *Ciência & Saúde Coletiva* 4 (2): 383–392.

Matthews, D. 2011. *Intellectual Property, Human Rights and Development: The Role of NGOs and Social Movements*. Cheltenham: Edward Elgar.

Mazzoleni R., and L.M.C. Póvoa. 2009. Accumulation of Technological Capabilities and Economic Development. Available at http://ideas.repec.org/f/pma824.html. Accessed 4 October 2016.

Mueller, B., and C. Mueller. 2014. *The Economics of the Brazilian Agricultural Development*. Working Paper, International Research Initiative on Brazil and Africa (IRIBA), University of Manchester, UK.

Nastari, P.M. 1983. *The Role of Sugar Cane in Brazil's History and Economy*. Retrospective Theses and Dissertations, n. 8948, Digital Repository of the Iowa State University.

Nehring, R. 2016. *Yields of Dreams: Marching West and the Politics of Scientific Knowledge in the Brazilian Agricultural Research Corporation (Embrapa)*. Paper No. 35 Presented at Global Governance Climate Justice and Agrarian Justice International Colloquium, 4–5 February 2016 (The Hague: The Netherlands).

Newell, P. 2008. Trade and Biotechnology in Latin America: Democratization, Contestation and the Politics of Mobilization. *Journal of Agrarian Change* 8 (2/3): 245–376.

Nunn, A. 2008. *The Politics and History of AIDS Treatment in Brazil*. New York: Springer.

Octaviani, A. 2010. Biotechnology in Brazil: Promoting Open Innovation. In *Access to Knowledge in Brazil*, ed. L. Sheaver, 79–102. London: Bloomsbury Academic.

OECD. 2001. *Using Knowledge for Development: The Brazilian Experience*. Paris: OECD Publishing.

Palma, J.G. 2012. *Brazil's Recent Growth*. UNCTAD and South Centre Discussion Paper No. 3 (Geneva: UNCTAD).

Pansera, C. 2016. Interview of Celso Pansera with Agência Brasil. Available at http://confap.org.br/news/brasil-vai-passar-de-pais-de-commodities-a-exportador-de-inovacao-diz-ministro/. Accessed 28 September 2016.

Paulani, L. 2010. Brazil in the Crisis of the Finance-led Regime of Accumulation. *Review of Radical Political Economics* 42 (3): 363–372.

Pedersen, J.D. 2008. *Globalization, Development and the State: The Performance of India and Brazil Since 1990*. Basingstoke: Palgrave Macmillan.

Peschard, K. 2012. Unexpected Discontent: Exploring New Developments in Brazil's Transgenics Controversy. *Canadian Journal of Development Studies* 33 (3): 326–337.

Philippou, S. 2005. Modernism and National Identity in Brazil, or How to Brew a Brazilian Stew. *National Identities* 7 (3): 245–264.

Pimentel, V.P., R.P. Gomes, T.L. Mitidieri, et al. 2014. *Inserção Internacional das Empresas Farmacêuticas: Motivações, Experiências e Propostas para o BNDES*. BNDES Report. Available at https://web.bndes.gov.br/bib/jspui/bitstream/1408/3107/1/BS%2040%20Inserção%20internacional%20das%20empresas%20farmacêuticas_P.pdf. Accessed 22 September 2016.

PITCE. 2003. Diretrizes de Política Industrial, Tecnológica e de Comércio Exterior, 26 de Novembro de 2003. Available at http://www.inovacao.unicamp.br/politicact/diretrizes-pi-031212.pdf. Accessed 10 September 2016.

Rada, N. 2013. Assessing Brazil's Cerrado Agricultural Miracle. *Food Policy* 38: 146–155.

Reis-Castro, L., and K. Henrickx. 2013. Winged Promises: Exploring the Discourse on Transgenic Mosquitoes in Brazil. *Technology in Society* 35: 118–128.

Reis, R., V. Terto, and C. Pimenta. 2009. *Intellectual Property Rights and Access to ARV Medicines: Civil Society Resistance in the Global South*. Rio de Janeiro: ABIA.

Ribeiro, L.L., and A.G. Carvalho. 2008. Private Equity and Venture Capital in an Emerging Economy: Evidence from Brazil. *Venture Capital* 10 (2): 111–126.

Rocha, G.M. 2002. Neo-Dependency in Brazil. *New Left Review* 16 (July–August): 5–33.

Rodriguez, A. 2008. *Knowledge and Innovation for Competitiveness in Brazil*. Washington, DC: World Bank Publications.

Santilli, J. 2012. *Agrobiodiversity and the Law: Regulating Genetic Resources, Food Security and Cultural Diversity*. New York: Earthscan.

Sauer, S., and S.P. Leite. 2012. Agrarian Structure, Foreign Investment in Land, and Land Prices in Brazil. *The Journal of Peasant Studies* 39 (3/4): 873–898.

Schwartz, R. 1988. Brazilian Culture. *New Left Review* 167: 77–90.

Schwartzman, S. 1995. *Science and Technology Policy in Brazil: A New Policy for a Global World*. Rio de Janeiro: Fundação Getúlio Vargas.

Scoones, I. 2005. *Contentious Politics, Contentious Knowledges: Mobilising against GM Crops in India, South Africa and Brazil*. Institute of Development Studies, Working Paper 256, Brighton UK.

Sell, S. 2003. *Private Power, Public Law: The Globalization of Intellectual Property Rights*. Cambridge: Cambridge University Press.

Shadlen, K. 2011. The Political Contradictions of Incremental Innovation: Lessons from Pharmaceutical Patent Examination in Brazil. *Politics & Society* 39 (2): 143–174.

Shadlen, K. 2017. *Global Change, National Responses: The New Politics of Pharmaceutical Patents in Latin America*. Oxford: Oxford University Press.

Shadlen, K., and E.M. Fonseca. 2013. Health Policy as Industrial Policy: Brazil in Comparative Perspective. *Politics and Society* 41 (4): 561–587.

Shaffer, G., M.R. Sanchez, and B. Rosenberg. 2008. The Trial at Winning at the WTO: What Lies Behind Brazil's Success. *Cornell International Law Journal* 41: 383–501.

Silva, R.E., A.A. Amato, and M.R.C.G. Novaes. 2015. Funding of Clinical Trials in Brazil for the Development of New Drugs: Who Are the Sponsors? *International Journal of Clinical Trials* 2 (4): 75–84.

Souza, V.S., and R.V. Santos. 2014. The Emergence of Human Population Genetics and Narratives About the Formation of the Brazilian Nation. *Studies in History and Philosophy of Biological and Biomedical Sciences* 47: 97–107.

Sweet, C. 2013. The Political Economy of Pharmaceutical Production in Brazil. In *The New Political Economy of Pharmaceuticals Production, Innovation and TRIPS*, ed. H. Löfgren. Basingstoke: Palgrave.

Tulchin, J.S., and H.A. Golding, eds. 2002. *Environment and Security in the Amazon Basin*. Washington, DC: Woodrow Wilson International Center for Scholars.

Velez E. 2010. *Brazil's Practical Experience with Access and Benefit Sharing and the Protection of Traditional Knowledge*. Policy Brief Number 8, ICTSD, Geneva, Switzerland.

Viana, A.L., H.P. Silva, and I. Yi. 2015. *Universalizing Health Care in Brazil: Opportunities and Challenges*. UNRISD Working Paper 2015–8. Geneva: UNRISD.

Watal, J. 2001. *Intellectual Property Rights in the WTO and Developing Countries*. The Hague: Kluwer Law International.

Welch, J.R. 2015. Brazil's New Biodiversity Law. *Ethnobiology Letters* 6 (1): 216–217.

Wilkinson, J., and S. Herrera. 2010. Biofuels in Brazil: Debates and Impacts. *The Journal of Peasant Studies* 37 (4): 749–768.

CHAPTER 7

Looking Ahead: The Emerging 'Knowledge Powers'?

Readers following the arguments presented so far will no doubt wish to raise a number of criticisms. One could be that the significance of the rupture occasioned by the shift towards the knowledge economy has been exaggerated in the case of India and Brazil. Data can be brought up to support this interpretation, for instance, the relatively low level of domestic R&D investment—currently around 0.8% GDP in India and 1.2% GDP in Brazil, compared with the world highest level of 4.15% GDP in South Korea—or the relatively low contribution high-tech sectors make to their economic performance: for example, high-tech exports as a percentage of manufacture exports in 2014 were only 8.5% and 10% for India and Brazil respectively, as compared with 27% for South Korea.[1] One way to interpret these figures is to argue that the shift towards the knowledge economy only concerned certain parts of the global economy, essentially the already economically and technologically advanced countries, or that the shift occasioned widespread pressures but that the Indian and the Brazilian states did not respond to them properly.

It is true that the knowledge economy—a socio-economic formation where wealth is predominantly generated from manipulating information and knowledge—is concentrated in the wealthier regions of the world economy, as is the fact that the shift towards it originating in these parts did not radically transform everything everywhere. In many ways, the shift

V. Muzaka, *Food, Health and the Knowledge Economy*, Building a Sustainable Political Economy: SPERI Research & Policy, DOI 10.1057/978-1-137-59306-1_7

is still ongoing as are the transformations in socio-economic forms that necessarily accompany it. More importantly perhaps, the shift towards this particular socio-economic formation, initially in the US and then in other wealthy countries that had a competitive advantage in the new technologies of the 1970s and 1980s, was not the natural result of the march of technology but rather was mediated politically by specific groups. In turn, the way this shift unfolded in practice was not uniform but the outcome of political conflicts occurring in contexts differentiated by their institutional peculiarities and the balance of political forces. Different and historically specific outcomes notwithstanding, against the backdrop of the real and politically constructed crisis of Fordism in key advanced countries, the knowledge economy imaginary did become dominant, as evidenced not only in the discourses and institutional reforms undertaken, but also in the growing contribution of the new technologies/sectors in the economic performance of leading OECD countries from the late 1970s onwards (Dutfield 2003: 195).

The main aim of situating the emergence of the knowledge economy imaginary in the shift towards post-Fordism in the first part of the book was to generate a theoretically informed account of why the socio-economic transformations accompanying this shift in key advanced countries were not limited to them. The adoption of the knowledge economy project and of the competition state orientation in key frontier states—and the attendant changes in state practices these moves occasioned—could not but alter the dynamics of the world economy and hence the field within which all states operate. Because states have always been spatially Janus faced—looking both inwards and outwards—and anchored simultaneously in the national and the global terrain, the inescapability of the shift towards the competition state orientation reaching states everywhere had as much to do with the nature of this shift as with the mutually constitutive nature of the national and the global: the nation-state and the world economy have developed together, or, put differently, the state made the world economy and the world economy made the state (Hobson 1997: 251). It is this structural and spatial feature of the modern state—yet to be claimed by another actor—that grants it, among other things, the capacity to make and remake its domestic socio-economic structures in response to changes in the world economy. This does not mean that the state functions as a filter for all external processes, but that in its continuous drive to adapt and respond simultaneously to internal and external pressures, it remains one of the central agents that shapes domestic and international institutional forms.

Not all states would have experienced the pressures emanating from the shift to the competition state orientation in the same manner, nor would they have been able to equally succeed in their attempts to transform their domestic socio-economic institutional forms to respond to these pressures. Everywhere the outcomes of institutional reforms initiated with a view to responding to the changing exigencies of international competition were and continue to be shaped both by the differently constituted domestic terrains where social conflicts unfold and by the specific manner in which domestic economies have been integrated in the global economy. The changes brought about by the embrace of the international competitiveness orientation on the part of key frontier economies are many, notable among them the renegotiation of the private–public, economic–non-economic and national–international boundaries in favour of private market players, increased commodification and dethroning the national economic space as the primary focus/source of growth. Discussing the manner in which the state in India and Brazil responded to these changes in the preceding analysis was not intended as a comparison between the two cases—their features are too different to succeed in this task—but rather as an investigation on how pressures generated by such changes found different expression in India and Brazil, shaped by the distinctive social, economic and political conditions domestically, as well as the relative position of each state in the world market.

Both the Indian and the Brazilian state were not only keenly aware of changes occurring in the global economy from the 1980s onwards but, as we have seen, contributed to them through their embrace of the knowledge economy imaginary and of the structural competitiveness orientation. That both got actively involved in the discursive and material constitution of the emergent knowledge economy imaginary and took the shift towards the competition state orientation seriously is visible in their official aim of becoming 'global knowledge powers' in the twenty-first century. This cannot be dismissed as mere rhetorical flourish; as developing countries with large domestic economies, long-nourished ambitions to catch up with frontier economies and distinct versions of exceptionalism, the Indian and the Brazilian states were perhaps more aware of the necessity to respond to the changing dynamics of the world economy than many other developing countries (China is a notable exception). We have seen that when for different reasons and constrained by specific structural conditions the state in each was contemplating a new growth strategy in the 1980s, the choice was made in favour of transforming economic and social structures towards achieving innovation-led structural competitiveness, although lower-value-added sectors were not excluded. Saying that a choice was made does not ignore the existence of pressures and constraining factors, but points to the taking of a particular position within a context that is not only (indeed, by definition, always) constraining, but also where alternative positions are possible.

In both India and Brazil the choice made was less due to direct pressure from outside groups or from important domestic interest groups than due to the adoption of the knowledge economy imaginary by key fragments within the state itself. That this choice was neither mere rhetoric nor a temporary experiment is visible in the undertaking of various institutional reforms that more often than not sought to emulate the institutional arrangements of what was perceived to be the most successful knowledge economy: the US. Although all growth regimes generate their share of social conflict, these tend to be more pronounced in periods when the coupling of the existing growth regime and its mode of socialisation breaks down and socio-economic transformations aimed at laying down the conditions necessary for a new growth regime are attempted. The outcomes of the institutional transformations attempted in India and Brazil with the aim of becoming knowledge economies do not resemble those which they sought to emulate nor could they do so, shaped as they were through social conflict unfolding in distinctive and differently constituted terrains. As noted at the start, the focus on key conflicts accompanying changes to the IP institutional form in India and Brazil was justified precisely on account of them offering a more interesting and revealing angle through which to explore the response of the Indian and Brazilian states to the exigencies of the new global knowledge economy and the way such response was expressed and contested in practice.

As the discussion on the Schumpeterian Workfare state (or the competition state) orientation in the second chapter revealed, the gamut of the reforms deemed necessary to respond to the exigencies of the knowledge economy is wide and obviously not limited to IP. An early sign of growing awareness regarding changes afoot in the world economy and of the urgent response required was the significant restructuring of the S&T orientation away from serving the needs of the domestic market towards improving international competitiveness in both India and Brazil in the 1980s. Like the Keynesian state in frontier economies, the state in India and Brazil had played a central role in establishing, financing and developing the domestic S&T base post-WWII; perceiving the shift towards the competition state orientation in key frontier states in the 1980s, it was the state in both that moved towards a new regime of growth based on international competitiveness and towards S&T reforms deemed necessary to succeed in the changed circumstances. Various existing S&T policies managed by different parts of the state were bought under the control of a newly established Science and Technology Department (India) and Ministry (Brazil) in 1985 with the aim of steering S&T investment towards transforming the existing base into one capable of supporting the new positive insertion in the world economy. It would be short-sighted to dismiss these moves as insignificant in light of the comparatively meagre public funds made available to support the new direction during this period. Relatively unimpressive R&D expenditure as a percentage of GDP says something about the macroeconomic situation in India and Brazil at the time, but it does not support an interpretation of the changes in S&T as being merely rhetorical.

Neither were these measures an instance of wishful thinking: in both India and Brazil, industrialisation efforts and investment in S&T post-WWII had by the early 1980s succeeded in creating a significant industrial and S&T base. It is true that the application of technology and the resulting degree of competitiveness varied greatly within and between their respective industrial bases, as is the fact that in both countries, considerable S&T investment had generated a rather uneven base marked by only a few 'pockets of excellence': atomic research, space technology and defence in the case of India, and aeronautics, agricultural research and petrochemicals in the case of Brazil. Even so, as in both cases industrialisation and S&T investment had been the linchpin of their strategy of catching up with the frontier economies, it was not an unthinkable step to peg on the existing industrial and S&T successes the new hopes of catching up with the frontier states when the latter embraced a growth strategy based on innovation and competitiveness in world markets. Gearing up towards this goal, the orientation of S&T policies shifted away from building a strong—and, in the case of India, self-sufficient—domestic technological base towards one that would support the competitiveness of their key sectors in world markets.

By virtue of constituting the pinnacles of the existing S&T structure and being under the direct control of the state, successful public research institutes and universities were among the first to experience the pressures of the new state orientation: as in key frontier states—the US and the UK in particular—public universities and research institutes were to play a key role in the knowledge economy project adopted by the Indian and the Brazilian state. We have seen how efforts on the part of the state to push public labs and universities towards a more active role in improving Brazil's competitive position had started since the late 1970s and resulted in many of them—notably FINEP and Embrapa—orienting their research towards technological innovation more directly relevant to Brazil's international competitiveness. Likewise, efforts to make public R&D more responsive to India's international competitiveness were most notable in radical changes to the work of the CSIR—then and now the largest public research institute in India—which from the early 1990s onwards shed its earlier responsibility of working towards improving indigenous technology towards making its research a key node in global R&D networks in tune with India's new project of becoming a competitive knowledge power.

Partly due to the new-found faith in market mechanisms and partly due to fiscal retrenchment, the new orientation was not to be achieved solely or even primarily by the state and public bodies but by the active engagement of private market actors. As had been the case with efforts to redraw the private–public boundaries in many frontier economies, the state in India and Brazil duly took measures to not only open more space for, but also to generously support and vigorously encourage, domestic businesses to boost their R&D activities in-house and in collaborations of various kinds. The most preferred of such collaborations were those between private companies and public research institutes and universities, the kind of partnerships that were believed to have been one of the key ingredients of success in the US especially after the 1980 Bayh–Dole Act. But responding to the ever-present and now more acute sense of a technological lag vis-à-vis frontier economies, partnerships with foreign companies were favoured as more important in achieving technological competitiveness. In this respect, the break with the past is rather notable, especially in the case of India. Having been historically more open than the Indian economy, the most dynamic sectors of the Brazilian economy had already been the object of FDI flows, which had had the effect of bringing many of them under foreign control by the 1980s without necessarily enhancing the technological dynamism of the Brazilian economy. Even so, as discussed, when FHC fully embraced the knowledge economy project and the international competitiveness orientation, FDI was seen as the best way of achieving technological upgrading and the doors were opened to foreign companies like never before. By contrast, the Indian state had severely restricted FDI from the mid-1960s onwards to a handful of sectors involving complex technologies; in line with its new orientation, it moved decidedly from the late 1980s onwards not only to grant a greater role to private domestic actors, but also to the entry of foreign companies as an important means of technological upgrading. In contrast to the earlier period, considerable incentives were given to Indian companies to cooperate with foreign companies, unencumbered by any kind of coordination or control over technological transfer. Unsurprisingly, the number of collaborations grew, but financial collaboration, especially in the form of M&As, dominated: while for every financial collaboration in 1977, there were 8 technological ones, by 2001 this number had fallen to 0.1 (Joseph and Abrol 2009: 113).

In yet another move inspired by institutional reforms undertaken in the US, both the Indian and the Brazilian state took measures to create and strengthen their domestic capital and especially venture markets, whose depth and health were believed to be strongly and positively related to the emergence and success of high-tech (especially life science) companies, which were expected to emerge as they had done in the US and elsewhere. As had been the case in the US early on, public funds were made available to support nascent venture capital markets, for example, through the Industrial Development Bank of India and the BNDES in Brazil. In addition, as we have seen, numerous state-funded programmes to support technological parks, business incubators and high-tech starts-ups were rolled out, accompanied by measures to entice institutional investors and private capital funds to get more involved in financing new and risky high-tech companies. In India, venture capital investment rules were liberalised from the mid-1990s onwards, while in Brazil, similar measures during this period were boosted by others aimed at enabling public institutions and pension funds to invest in venture capital vehicles.

It is in the context of these institutional reforms motivated by the knowledge economy having become a state project in India and Brazil that changes to their respective IP regimes are to be located. Because of IP's centrality to the knowledge economy, because of its effects on the distribution of social wealth and because of the nature of the pharmaceutical and biotech sectors in each country, IP reforms occasioned multiple and ongoing social conflicts in India and Brazil. As discussed in Chaps. 5 and 6, the specific shape the domestic IP regime for pharmaceutical and plant genetic resources took in practice bears the distinctive imprints left by such conflicts. The state was at once the locus towards which these conflicts gravitated and the main actor participating in them. Responding to the new obligations undertaken in TRIPS, the state in both India and

Brazil was the main architect of IP reforms. Equally important, intent on transforming socio-economic structures in order to compete in the world market and win, the state was often the keenest advocate of strong IP protection levels. In Brazil, it succeeded in offering early and rather generous IP protection for pharmaceuticals until, amidst mounting conflicts, its inability to balance growing rents transferred to foreign pharmaceutical companies and its constitutional duty towards public health care caused it to temper somewhat its zeal. In India, the success of the generics sector that the state had nourished for years and a strong coalition that was simultaneously pro-Indian business and pro-health generated a pharmaceutical IP regime equipped with more 'defensive' measures, but the state is likely to seek to repeal them if and when they are perceived to hinder its goal of transforming India's pharmaceutical sector into a globally innovative one. It has been this same drive that has marked the state's approach towards IP protection for plant genetic resources and biotech in both India and Brazil. As the respective biotech sectors were still rather underdeveloped at the time, and given the role IP titles had played in the commercialisation of biotech in the US, the state in India and Brazil appeared convinced that the success of its biotech and agro-biotech sectors necessitated strong IP protection levels, an orientation that brought it in conflict with a considerable number of civil society groups.

Despite differences, it could be confidently said that in guiding IP reform in India and Brazil, the state was not reluctantly responding to international IP obligations it had no choice but to accept, but rather it was seeking to transform domestic socio-economic structures in line with the new competition and knowledge economy orientation it had adopted. Of course, IP titles do not generate knowledge—people do—and much less the broader non-economic resources on which innovation relies. Nevertheless, IP is the key means through which various kinds of social knowledge and non-economic resources can be commodified, appropriated and put in the service of wealth generation. This is one of the reasons why the state in both India and Brazil often appeared to be the keenest domestic advocate of strong IP protection levels. That this position is fully in line with the kind of knowledge economy imaginary it embraced is visible in the key role the state in both has played not only as a knowledge market creator, but also as the most important player in domestic knowledge markets: state agencies and public research institutes hold the largest share of domestic IP assets. This phenomenon is understandable in light of the new competitiveness orientation and of the fact that those parts of the economy that were best positioned to participate in the knowledge economy were largely public, thanks to the considerable investments both states had made earlier in building and strengthening their domestic scientific and research infrastructure.

Even so, the case remains that, helped in no small measure by the IP reforms undertaken, the majority of IP titles granted domestically in India and Brazil go to foreign companies, which use them not to improve the technological capabilities of the respective sectors, but rather to expand and extend the technological rents they are able to extract from such titles. Despite years of reforms and efforts to improve ranking in international patent league tables, the Indian and Brazilian economies are still net IP importers and not insignificant generators of rents for (largely foreign) IP-holders. A shortcoming that both the Indian and the Brazilian state had displayed in their earlier developmental period—namely their inability to control the 'rent-seeking' behaviour of private domestic business—now appears to have been made worse by their inability or unwillingness to control the rent-seeking behaviour of foreign businesses that operate in the key high-tech sectors of the global knowledge economy. It is often protection against this behaviour that state agencies in both India and Brazil invoke as underpinning their own accumulation of IP titles domestically. In what manner and towards what ends these domestic and nominally public IP assets will be used remains an open question. The signs so far are not particularly encouraging. Having exchanged its role of undertaking R&D towards improving indigenous technology for a market-oriented mission, the CSIR has accumulated the largest number of domestic patents in India and many abroad, but only a small part of them have been licensed, or the technologies they protect been transferred domestically (Joseph and Abrol 2009; Abrol 2013). A similar parallel can be drawn with Embrapa in Brazil, the public agricultural research agency whose success resembles that of the CSIR in India. It, too, has accumulated a large number of IP titles domestically, but its new mission of commercially exploiting the outcome of publically funded research has been geared towards improving Brazil's agro-biotech competitiveness in world markets, limiting in the process, as we saw, the rights and fortunes of small Brazilian farmers.

Some of these arguments may appear to be at odds with the position that has characterised the involvement of the Indian and the Brazilian state in contests over IP in various international fora. Since the entry in force of the TRIPS Agreement in 1995, both have led a number of other developing countries in challenging various aspects of the global IP regime, for example, concerning the restrictive impacts of IP on access to affordable medicines, on the protection of genetic biodiversity and on the necessary policy space to design relevant technological strategies. This is not necessarily surprising. Although this is more pronounced in the case of India, both states have an ambivalent attitude towards the global IP regime: on the one hand, being ambitious catch-up contenders and net IP-importers, they experience the global IP rules as loaded against them, while on the other, their embrace of the logic of competitiveness and knowledge economy imaginary made transforming their domestic IP regime in line with these very rules not only critical but seemingly indispensable for achieving their global ambitions. If contests in the global IP regime have intensified and India and Brazil are key among players involved in them, this has less to do with these state actors attempting to challenge the 'rules of the game' than with attempting, as other state actors do, to expand the benefits that can be gained from playing the game through shaping the rules in this particular policy regime in ways that favour them. In other words, given the current orientation of the Indian and the Brazilian state, there can be little hope in the radical transformation of the global IP regime coming from their leadership of other developing countries. Not only in this regime but in other international regimes, too, India's and Brazil's increased participation being interpreted as an effort on their part to challenge or rewrite the 'rules of the game' is both exaggerated and misconstrued.

That neither India nor Brazil has any serious intention of radically upsetting the rules underpinning the current economic order—except emerging at its apex—is clear in the state orientation they adopted in the 1980s. Their project of becoming global knowledge powerhouses is unmistakeably unoriginal: in their drive to catch up and compete in the global knowledge economy, neither India nor Brazil has been particularly innovative, borrowing this imaginary, alongside the spirit and often the form of institutional arrangements, from advanced economies, especially the US. A uniquely Indian or Brazilian knowledge economy imaginary has not been forthcoming. If the autonomy of all states is constrained to varying degrees by their position in the world economy, this constraining effect must be even more pronounced for those that, like India and Brazil, are not yet located at its core. The more these relative newcomers got drawn into the system of states and into the world economy, the more constrained they became: in seeking to replicate the form of Western nation-states and catch up with them economically, possibilities of unique social–economic–political identities were abandoned long ago, a development which further restrains their space of experimenting with alternative forms of being and operating in the world economy today.

Nevertheless, their respective knowledge economy projects are necessarily locally refracted. One of the characteristics that makes the Indian and the Brazilian knowledge economy projects different from others is the manner in which hopes of becoming 'knowledge powers' have been co-articulated and fused with different and sometimes conflicting versions of nationalism, versions that are always changing and ever amenable for use by political actors. Even if nationalism(s) were not successfully mobilised earlier by the Indian and the Brazilian state to create a solidaristic national project—and the vast energy needed to achieve it—when compared with the East Asian developmental states, both systematically exploited nationalism to generate popular support and legitimation for their respective development strategies during most of the twentieth century. It is therefore not surprising that the new growth strategy at the turn of this century should also be legitimised on a nationalist imaginary of transforming India and Brazil into 'knowledge powers' competing and succeeding in the global market. This popular legitimising imaginary is also underpinned and further enhanced by their distinct versions of exceptionalism: *re*gaining its world power status in the case of India, and finally reaching the great power status its size and natural wealth had bestowed it in the case of Brazil. For a growth strategy based on innovation and structural competitiveness, for which even-higher levels of societal and economic mobilisation are required, a popular and hopeful imaginary is correspondingly more indispensable.

If no doubts about the embrace of the knowledge economy imaginary as a state project on the part of the Indian and the Brazilian state remain, their performance to date mentioned at the start of this chapter can be seen to represent a case of insufficient progress that necessitates additional and further-reaching reforms. Most reports generated by international observers and commentators, for example, the OECD, *The Economist* or international business consultancies, are along these lines. But the conceptual framework and analysis offered here invite a deeper investigation into how the tensions characterising post-Fordism, alongside those emanating from the specific position of the Indian and Brazilian economies in the global economy, ought to be taken into account when evaluating their performance. Of the various lines of enquiry of this kind that could be followed in the future, one would necessarily focus on the enormous difficulty *any* state—developing or otherwise—faces in simultaneously having to invest in education, research infrastructure, health care and other elements contributing to the long-term (re)generation of collective social knowledge and the non-economic realm on the one hand, and in sustaining the conditions for subordinating these elements to commodification in the service of private wealth generation on the other. These are not new tensions, but have become more challenging for at least three reasons: first, because the knowledge economy, as we have seen, relies on a deeper and more intensive commodification and appropriation of collective social knowledge than other formations; second, because despite the new-found faith in markets, private returns to long-term investments of the kind mention above would never be high or immediate enough to motivate the private sector to provide them; and, third, because the state must finance these ever-higher investments under conditions of fiscal retrenchment.

Adding to these challenges are others specific to the political economy of India and Brazil. The complexities of teasing these out not only are beyond the scope of this work, but would also require substantially more space than is available. Still, it is possible to sketch out some of the main challenges India and Brazil face in common, namely becoming competitive knowledge economies in the face of de-industrialisation, financialisation and daunting domestic inequality levels. Despite accounts circulating in many advanced economies that depict blue-collar work having migrated to developing countries, de-industrialisation is a phenomenon that has affected both developed and developing countries. The debate over its sources and effects continues, and technological change is often named as the main culprit. Indeed, if data is to be believed, the 'softening' and 'dematerialisation' of the economy as a result of the Digital Revolution is visible in rising volumes of knowledge-intensive trade, that is, trade in goods and services where R&D or skilled labour account for the larger share of value. Such volume was said to have reached US$12.6 trillion in 2012, more or less half the total value of trade in goods, services and finance.[2] Whatever is made of these figures, it is certainly true that this process has not been limited to advanced economies alone. Both the Indian and the Brazilian economies have been de-industrialising for some time: the share of employment engaged in manufacturing in Brazil peaked in the 1980s at 15% of total employment, whereas in India, it peaked later (2002) but at the even-lower rate of 13% (Rodrik 2013). The rise of China was not the only cause: its own manufacturing share of employment peaked at 15% in the mid-1990s, again at a much lower level compared with a share of around 35% maintained for over a century in England, the original 'factory of the world' (Evans 2008).

De-industrialisation poses difficulties to all developing countries because it has undercut the growth strategy that historically underpinned the success of the advanced economies of today. It is rather more damaging for them because their de-industrialisation was not the result of their economies having reached 'maturity': with the exception of a few Asian countries, many of them have experienced falling manufacturing shares in both employment and real value-added at levels of income that are a fraction of those at which the advanced economies started to de-industrialise (Rodrik 2013, 2015). One of the challenges premature industrialisation introduces, in turn, is that sectors that 'replace' manufacturing—notably, services and commodities in the case of Brazil and services and *informal* manufacturing in the case of India—are unlikely to be of the high-value-added or technologically advanced kind (Tregenna 2016). If growth in the future is to come predominantly from services, being caught in this dynamic is likely to pose similar growth and competitiveness challenges like those faced by developing countries unable to move up high-tech and higher-value-added manufacturing sectors in the previous growth regime.

Services and whatever manufacturing remains in India and Brazil is likely to be predominantly of a low-value-added type partly because de-industrialisation occurred prematurely; that is, industrialisation had not yet succeeded in creating a cohesive domestic productive structure characterised by strong links between well-developed downstream industries and the capital-good, high-tech upstream ones. The existence of a number of 'pockets of excellence' cannot fill the gap left by a much diminished and technologically weaker manufacturing base in the wake of premature de-industrialisation. As was discussed in the preceding chapters, the industrial base in both India and Brazil by the 1980s was not notable for its high levels of technological dynamism; besides, India and particularly Brazil would also see key components of their domestic productive and innovation systems uprooted and further weakened during the late 1980s and 1990s. This leaves both at a disadvantageous position as far as succeeding in the knowledge economy of the twenty-first century is concerned. This is so not only because their manufacturing base is unlikely to push productivity and economic growth forward as things stand, but also because, being traditionally closely linked to technological innovation, their current state is unlikely to generate the necessary capabilities to operate close to the technological frontier, a prospect made even more difficult by the existence of a rather restrictive global IP regime.

It is important at this juncture to underline that the de-industrialisation of the kind visited upon India and Brazil was not the outcome of the march of technology. Rather, it was the outcome of the Indian and the Brazilian state responding to the changed dynamics of the world economy by embracing the competitive state orientation and the attendant shift of focus away from the domestic to the global economy as the main source of growth. As they opened up their economies—more cautiously, but still decisively, in the case of India—and deepened their integration with other economies enjoying much higher levels of economic and technological sophistication, their most advanced industrial sectors were the first casualties. It is in this sense that India and Brazil can be said to have 'imported' de-industrialisation from other countries (Rodrik 2015; Tregenna 2016), but this was not entirely forced on them. While their dual anchorage in the national and the global economy necessarily limits the states' room for manoeuvre, we have seen that it were key fragments within the state in each case that decided to respond to the changes in the world economy during the 1980s in favour of the knowledge economy and competitive state orientation. The continental size of the states' economies—India's more so than Brazil's—offered other options, for example, a hybrid, inward–outward-looking development model akin to China's (Boyer 2016), which was simply not available to smaller and less developed countries.

But neither was technology the main culprit for de-industrialisation in advanced economies. To be sure, technological change of the kind ushered in through the Digital Revolution would necessarily cause changes to the domestic productive fabric, but there is no reason to believe these should result in its weakening. As was discussed in the second chapter, as key advanced economies—the US and the UK in particular—embraced the competitive state orientation and undertook a number of measures that radically redrew the boundaries between the national–international and public–private in favour of private market actors and of international finance, it gradually became evident that leaving it to private market actors to define their own financial, industrial and technological strategies in the context of increasingly liberalised and internationalised financial flows tended to subordinate such strategies to short-term and often purely financial goals (Chesnais 1991; Weiss 2014). The impact of such shifts on domestic productive structures has been significant—among others, the loss of manufacturing jobs, rising inequality and decline in innovative capacity (Rodrik 2015)—and not limited to these countries. We also saw that, liberated from national controls, ever-rising financial flows came to colonise the most high-tech parts of the global economy without making them more dynamic. More often than not, the deepening of capital markets meant not an increase in available funds to support R&D, but the encroachment of short-termism and shareholder value on corporate planning and investment practices: the collapse of the average holding period for stocks has been so profound that it makes little sense to speak of long-term stockholders anymore (Crotty 1990).

Unsurprisingly, companies spend a considerable share of their profits in stock buybacks—around US$3 trillion by S&P500 companies in the last decade—as they find that market gimmicks of this kind are more rewarding than long-term investment (Mazzucato 2015). R&D investment has suffered especially: to make but one example from the life sciences, during 1992–2011, Amgen's[3] stock repurchases surpassed its R&D expenditures every year, except 2004 (ibid.). Generally speaking, the short-term and rentier nature of financial capital markets and flows contrasts starkly with the stable, considerable and rather long-term investment required to build innovative productive capabilities. This tendency should have already been a cause for alarm for Indian and Brazilian policymakers, who, inspired by the (US) received wisdom that stock market speculation is necessary for innovation, have taken measures to develop their own versions of venture capital markets for high-tech companies. But as was discussed with reference to the emergence of biotechnology in the US in the third chapter, the largest share of investment in R&D in this and other new technologies was shouldered by public funds, which are typically much more longer term and risk-taking than venture capital (Block 2011; Weiss 2014; Mazzucato 2015). It seems reasonable to suggest that, as the increasing weight of finance on the economy negatively affects the level of technological sophistication of the domestic productive base, one of the first steps to be taken by India and especially Brazil (given the much higher level of financialisation of its economy), would be to find ways to de-financialise and make finance primarily national.

Under continued financialisation, the challenges of India and Brazil becoming innovative knowledge-based economies in the absence of a strong domestic productive base are colossal. The record of this sort of growth regime in advanced economies with more sophisticated productive and innovative systems than theirs has so far been characterised by low growth rates and increased inequalities (Coriat and Schméder 2006). Hopes pegged on services to deliver growth and competitiveness in the future could disappoint: if they remain of the low-skilled and low-value-added kind, they will likely perpetuate the spiral of low growth, stagnant wages and growing inequalities (Evans 2008). Growing inequality levels are a problem everywhere, but particularly so for India and Brazil, and indeed for other developing countries, where industrialisation strategies imitating the Keynesian state orientation of advanced economies post-WWII were never accompanied by similar social, distributional and welfare policies. Grotesque inequality and socio-economic marginalisation levels are a persistent feature of the Brazilian political economy, and India has not managed to buck the trend: at just under 0.4, still lower than over 0.5 of the wealthier Brazil,[4] India's Gini coefficient is considerable and still growing. High inequality levels in both India and Brazil are made worse by the existence of large informal sectors: the percentage of people in informal employment, that is, in jobs in the informal and formal sectors—excluding agriculture—who do not enjoy basic social or legal protection was 42.2% for Brazil and 83.6% for India in 2009 (ILO 2012).

High inequality and socio-economic marginalisation levels are first and foremost morally unacceptable, but they also hinder further progress, certainly of the kind on which India and Brazil have fixed their sights. A narrow focus on technology and innovation will not deliver for all the reasons mentioned above, as well as the simple but often overlooked fact that the social, economic and technological order are mutually shaped and constituted. High levels of socio-economic marginalisation in practice materialise in unequal and insufficient access to nutrition and health care—a basic precondition for life and work—unequal access to education, insufficient work opportunities, poor working conditions, inadequate social protections and so on. The relative neglect of investments in general education, for instance, means that India is today one of the largest single-country contributors to the pool of illiterate people in the world, with a total adult illiteracy rate of 37%—Brazil's is 10%—while poor quality hampers in different ways the primary and secondary education systems in both countries.[5] Incidentally, this is one of the reasons, among many others, why remarkable successes in certain 'pockets of excellence' remain localised and have not succeeded in generating enough forward and backward linkages to push the Indian and Brazilian economies towards higher levels of technological dynamism. Likewise, despite impressive improvements through the Zero Hunger Programme, around 30.2% of households were in some degree of food insecurity in 2010 in Brazil, the world's fourth-largest food exporter (Chmielewska and Souza 2011). The situation in India is considerably worse: during 2000–2007, nearly half of its under-five children were malnourished, worse than the worst performer in the African region (Chopra 2011). As regards health care, we saw that despite the historical rupture occasioned by the *movimento sanitário* with the introduction of a universal national healthcare system (SUS) in Brazil in 1988, the system's chronic underfunding means that more than 70% of Brazilians who rely on it spend less than or nearly as much as the 30% of the population who relies on private healthcare provisions (Muzaka 2017). Only in 2015 did the government draft a policy recommending the creation of a universal healthcare system in India, where the private providers cover nearly 80% of outpatient and about 60% of inpatient care for those who can afford it (GoI 2015).

If services are to be the main source of growth in the future, they are unlikely to improve these indicators, unless they are of the high-tech and high-value-added kind. As growth based on low-skilled services tends to exacerbate inequalities, the need for welfare and distributional policies will be more rather than less pronounced than before. In the face of such high levels of socio-economic marginalisation in both countries, the path to becoming knowledge economies requires institutional innovations that differ from what has so far been attempted. The purpose of the state managing competition in world markets cannot be simply that of increasing the economic benefits accruing from the processes of internationalisation, but should primarily be that of managing these processes so as to protect the domestic population from their worst effects. A different kind of a knowledge economy imaginary could be created towards this end. Instead of orienting reforms and investments towards preparing and turning ever-larger parts of their societies over to the competitive treadmill of the global economy, India's and Brazil's plentiful social and economic challenges could instead provide the basis for an alternative knowledge economy imaginary in which meeting the numerous and often basic needs of the many becomes simultaneously the means and the ends. At the very least, a knowledge economy based on social needs will necessitate a set of institutional innovations that introduce more social protections and distributional policies, a technological policy that counters the tendency of capital to generate innovations based on short-term returns regardless of social or environmental costs towards meeting social needs, and an economic orientation with an efficient division of labour, where goods are homespun when possible and finance is primarily national.[6] Because competitiveness is based on the robustness of the domestic structures—social and economic—a focus on what goes on at home is even more justified. India's and Brazil's continental-sized economies offer them an opportunity to reorient themselves in this direction. Whether the Brazilian and the Indian state will contemplate such imaginaries and succeed in realising them remains to be seen. It may be that whatever else the incipient protectionism in frontier economies promises—protectionism of the type embodied in the current Trump administration in the US—it may also offer countries such as India and Brazil more space to experiment with different imaginaries, should they wish to.

Notes

1. Data from the World Bank database: http://data.worldbank.org/indicator/GB.XPD.RSDV.GD.ZS (for R&D data) and http://data.worldbank.org/indicator/TX.VAL.TECH.MF.ZS (for export data), both accessed on 25 November 2016.
2. The Economist (2014) 'Emerging Economies: Arrested Development', 4th October, available at http://www.economist.com/news/special-report/21621158-model-development-through-industrialisation-its-way-out-arrested-development, accessed on 25 November 2016.
3. The example of Amgen (est. 1980 in the US) is interesting also because it is currently the largest biotech company in the world.
4. Taking into account purchasing power differences, per capita GDP in Brazil is almost three times that in India; Gini coefficients quoted here (for 2011) are, respectively, the Gini coefficient of income for Brazil and of expenditure for India (see data from Cebrap, Brazil, available at http://cebrap.org.br/v3/arquivos/pesquisas/principais-resultados-da-pesquisa-labour-market-inequality-in-brazil-and-india-4917.pdf, last accessed on 26 November 2016).
5. Data from UNICEF databases https://www.unicef.org/infobycountry/india_statistics.html (India) and https://www.unicef.org/infobycountry/brazil_statistics.html (Brazil), both accessed on 25 November 2016.
6. This is a reference from Keynes' 'let goods be homespun whenever it is reasonably and conveniently possible, and, above all, let finance be primarily national' in his essay 'National Self-Sufficiency' (1933), available at https://www.mtholyoke.edu/acad/intrel/interwar/keynes.htm, last accessed on 5 April 2017.

Bibliography

Abrol, D. 2013. Where Is India's Innovation Policy Headed? *The Social Scientist* 41 (3–4): 65–80.

Block, F. 2011. Crisis and Renewal: The Outlines of a 21st Century New Deal. *Socio-Economic Review* 9: 31–57.

Boyer, R. 2016. Development and Regulation Theory. In *Handbook of Alternative Theories of Economic Development*, ed. E. Reinert, J. Gosh, and R. Kattel, 352–385. Cheltenham: Edward Elgar.

Chesnais, F. 1991. Technological Competitiveness Considered as a Form of Structural Competitiveness. In *Technology and National Competitiveness: Oligopoly, Technological Innovation, and International Competition*, ed. J. Niosi, 142–176. Montreal: McGill-Queen's University Press.

Chmielewska, D., and D. Souza. 2011. The Food Security Context in Brazil: Country Profile. International Policy Centre for Inclusive Growth (UNDP). Available at http://www.ipc-undp.org/pub/IPCCountryStudy22.pdf. Accessed 6 April 2017.

Chopra, R. 2011. Calorie Intake of the Rural Poor Has Drastically Fallen, Says Planning Commission Report. *India Today*, October 22. Available at http://indiatoday.intoday.in/story/poverty-malnutrition-india-hunger-index/1/157110.html. Accessed 6 April 2017.

Coriat, P., and G. Schméder. 2006. Post-Fordism in a More Globalized Capitalism. In *The Hardship of Nations: Exploring the Paths of Modern Capitalism*, ed. P. Coriat, P. Petit, and G. Schméder, 311–340. Cheltenham: Edward Elgar.

Crotty, J.R. 1990. Owner-Manager Conflict and Financial Theories of Investment Instability: A Critical Assessment of Keynes Tobin and Minsky. *Journal of Post-Keynesian Economics* 12 (4): 519–542.

Dutfield, G. 2003. *Intellectual Property and the Life Sciences Industries: A Twentieth Century History*. Aldershot: Dartmouth Publishing.

Evans, P. 2008. *In Search of the 21st Century Developmental State*. Working Paper no. 8, Centre for Global Political Economy, University of Sussex, UK.

GoI. 2015. *Draft National Health Policy*. Ministry of Health & Family Welfare. Available at http://www.mohfw.nic.in/showfile.php?lid=3014. Accessed 20 February 2017.

Hobson, J. 1997. *The Wealth of States: A Comparative Sociology of International Economic and Political Change*. Cambridge: Cambridge University Press.

ILO. 2012. Statistical Update on Employment in the Informal Economy, ILO – Department of Statistics. Available at http://laborsta.ilo.org/applv8/data/INFORMAL_ECONOMY/2012-06-Statistical%20update%20-%20v2.pdf. Accessed 6 April 2017.

Joseph, K.J., and D. Abrol. 2009. Science, Technology and Innovation Policies in India: Achievements and Limits. In *BRICS and Development Alternatives: Innovation Systems and Policies*, ed. J.E. Cassiolato and V. Vitorino, 101–131. London: Anthem Press.

Mazzucato, M. 2015. *The Entrepreneurial State: Debunking Public vs. Private Sector Myths*. London: Anthem Press.

Muzaka, V. 2017. Lessons from Brazil: On the Difficulties of Building a Universal Healthcare System. *The Journal of Global Health* 7 (1): 010303.

Rodrik, D. 2013. The Perils of Premature Deindustrialization. *Project Syndicate*, October 11.

———. 2015. *Premature Deindustrialization*. Working Paper No. 20935. Cambridge, MA: National Bureau of Economic Research.

Tregenna, F. 2016. Deindustrialisation and Premature Deindustrialisation. In *Handbook of Alternative Theories of Economic Development*, ed. E. Reinert, J. Gosh, and R. Kattel, 710–728. Cheltenham: Edward Elgar.

Weiss, L. 2014. *America Inc.* Ithaca, NY: Cornell University Press.

Bibliography

Abbott, F. 2009. TRIPS Dispute Settlement Decisions. Presented at International Centre for Trade and Sustainable Development (ICTSD) Seminar on 26–27 October 2009, Switzerland. Available at http://www.ictsd.org/downloads/2009/10/abbott_trips_dsu.pdf. Accessed 5 May 2016.

Achilladelis, B. 1999. Innovation in the Pharmaceutical Industry. In *Pharmaceutical Innovation – Revolutionizing Human Health*, ed. R. Landau, B. Achilladelis, and A. Scriabine. Philadelphia: Chemical Heritage Press.

Abrol, D. 2013. Where Is India's Innovation Policy Headed? *The Social Scientist* 41 (3–4): 65–80.

Ackerman, F. 1971. Industry and Imperialism in Brazil. *Review of Radical Political Economics* 3: 1–39.

Adede, A.O. 2003. Origins and History of the TRIPs Negotiations. In *Trading in Knowledge: Development Perspectives on TRIPs, Trade and Sustainability*, ed. C. Bellmann, G. Dutfield, and R. Melendez-Ortiz, 23–35. London and New York: Earthscan.

Alamgir, J. 2009. *India's Open-Economy Policy: Globalism, Rivalry, Continuity.* Abingdon: Routledge.

Anderson, P. 2011. Lula's Brazil. *London Review of Books*. Last accessed at http://www.lrb.co.uk/v33/n07/perry-anderson/lulas-brazil. Accessed 2 October 2016.

Angell, M. 2004. *The Truth About the Drug Companies: How They Deceive Us and What to Do About It*. New York: Random House.

Anuradha, R.V., B. Taneja, and A. Kothari. 2001. 'Experiences with biodiversity Policy-Making and Community Registers in India. In *Participation in Access*

V. Muzaka, *Food, Health and the Knowledge Economy*, Building a Sustainable Political Economy: SPERI Research & Policy, DOI 10.1057/978-1-137-59306-1

and Benefit-Sharing Policy, ed. K. Swiderska, 1–59. London: International Institute for Environment and Development (IIED).

Balbachevsky, E. and A.J.J. Botelho 2011. Science and Innovation Policies in Brazil: A Framework for the Analysis of Change and Continuity. Paper Presented at the IPSA-ECPR Joint Conference: Whatever Happened to North-South? São Paulo, Brazil, February 2011.

Ballance, R., J. Pogany, et al. 1992. *The Worlds Pharmaceutical Industries: An International Perspective on Innovation, Competition and Policy*. Aldershot: Edward Elgar.

Barros, R., M. de Carvalho, S. Franco, and R. Mendoça. 2010. Markets, the State and the Dynamics of Inequality in Brazil. In *Declining Inequality in Latin America: A Decade of Progress?* ed. L.F. Lopez-Calva and N.C. Lustig. Baltimore, MD: Brookings Institution Press.

Basheer, S. 2005. Policy Style Reasoning at the Indian Patent Office. *Intellectual Property Quarterly* 3: 309–323.

———. 2009. Encouraging Drug Innovation. *LiveMint*, August 27. Available at http://www.livemint.com/Opinion/v4H58mw6oSVDJOl2hKkWbK/Encouraging-drug-innovation.html. Accessed 5 July 2016.

Becker, J., J. Jäger, B. Leubolt, and R. Weissenbacher. 2010. Peripheral Financialization and Vulnerability to Crisis: A Regulationist Perspective. *Competition and Change* 14 (3–4): 225–247.

Bergeron, S. 2004. *Fragments of Development: Nation, Gender, and the Space of Modernity*. Ann Arbor: University of Michigan Press.

Berman, E. 2014. Not Just Neoliberalism: Economization in US Science and Technology Policy. *Science, Technology & Human Values* 39 (3): 397–431.

Bhagwati, J. 1993. *India in Transition: Freeing the Economy*. Oxford: Clarendon Press.

Bhattacharjea, A., and F. Sindhwani 2014. *Competition Issues in the Indian Pharmaceuticals Sector*. Working Paper, Centre for Development Economics, Delhi School of Economics, New Delhi, India. Available at http://www.cuts-ccier.org/compeg/pdf/report-pharmaceutical_sector_study.pdf. Accessed 20 May 2016.

Biehl, J. 2004. The Activist State: Global Pharmaceuticals, AIDS and Citizenship in Brazil. *Social Text* 22 (3): 105–132.

BioMinas. 2011. *The Brazilian Life Science Industry: Pathways for Growth*. Report available at http://biominas.org.br/en/downloads/the-brazilian-life-science-industry-pathways-for-growth-biominas-e-pwc/. Accessed 20 September 2016.

Birch, K., and D. Tyfield. 2012. Theorizing the Bioeconomy: Biovalue, Biocapital, Bioeconomics or … What? *Science, Technology & Human Values* 38 (3): 299–327.

Block, F. 2011. Crisis and Renewal: The Outlines of a 21st Century New Deal. *Socio-Economic Review* 9: 31–57.

Borrás, S. 2008. Innovation Policy and Institutional Competitiveness in Europe and Denmark. In *Institutions and Politics*, ed. P. Nedergaard and J.L. Campbell, 53–72. Copenhagen: DJØF Publishing.

Borrus, M. 2002. *The Global Pharmaceutical Market: International Trade and Competition in High Technology*, University of California, Haas Business School, Berkeley.

Borup, M., N. Brown, K. Konrad, and H. Van Lente. 2006. The Sociology of Expectations in Science and Technology. *Technology Analysis & Strategic Management* 18 (3–4): 285–298.

Bose, I. 2010. *How Did the Indian Forest Rights Act 2006 Emerge?* IPPG Discussion Papers, No. 39. University of Manchester. Available at http://re.indiaenvironmentportal.org.in/files/forest%20rights%20act%202006-emerge.pdf. Accessed 15 July 2016.

Bound, K. 2007. *India, the Uneven Innovator*. London: Demos.

———. 2008. *Brazil: The Natural Knowledge Economy*. London: Demos.

Bourdieu, P. 1987. The Force of Law: Towards a Sociology of the Juridical Field. *Hastings Law Journal* 38 (July): 805–853.

Boyer, R. 2016. Development and Regulation Theory. In *Handbook of Alternative Theories of Economic Development*, ed. E. Reinert, J. Gosh, and R. Kattel, 352–385. Cheltenham: Edward Elgar.

Boyd, W., S. Prudham, and A. Schurman. 2010. Industrial Dynamics and the Problem of Nature. *Society & Natural Resources: An International Journal* 14 (7): 555–570.

Boyer, R. 1990. *The Regulation School: A Critical Introduction*. New York: Columbia University Press.

Boyle, J. 2003. The Second Enclosure Movement and the Construction of the Public Domain. *Law and Contemporary Problems* 66: 33–74.

Brand, U., and C. Görg. 2013. Regimes in Global Environmental Governance and the Internationalization of the State: The Case of Biodiversity Politics. *International Journal of Social Science Studies* 1 (1): 110–122.

Brand, U., C. Görg, J. Hirsch, and M. Wissen. 2008. *Conflicts in Environmental Regulation and the Internationalisation of the State: Contested Terrains*. London: Routledge.

Brenner, N., B. Jessop, M. Jones, and G. MacLeod. 2008. State Space in Question. In *State/Space: A Reader*, ed. N. Brenner, B. Jessop, M. Jones, and G. MacLeod, 1–26. Hoboken, NJ: Wiley.

Brown, N. 2003. Hope Against Hype – Accountability in Biopasts, Presents and Futures. *Science Studies* 16 (2): 3–21.

Bryan, D., M. Rafferty, and D. Wigan. 2017. Capital Unchained: Finance, Intangible Assets and the Double Life of Capital in the Offshore World. *Review of International Political Economy* 24 (1): 56–86.

Caliari, T., and R.M. Ruiz. 2014. Brazilian Pharmaceutical Industry and Generic Drugs Policy: Impacts on Structure and Innovation and Recent Developments. *Science and Public Policy* 41 (2): 245–256.

Carvalho, J.M. 2000. Dreams Come Untrue. *Daedalus* 129 (2): 57–82.

Casper, S. 2007. The German Biotechnology Industry: The Limits of Orchestrated Innovation. In *Creating Silicon Valley in Europe: Public Policy Towards New Technology Industries*, ed. S. Casper, 74–104. Oxford: Oxford University Press.

Cassiolato, J.E., and H. Schmitz, eds. 1992. *Hi-Tech for Industrial Development: Lessons from the Brazilian Experience in Electronics and Automation*. London: Routledge.

Cassiolato, J.E., G.F. Zucoloto, M.S. Rapini, and S.G.A. Souza. 2011. The Recent Evolution of the Biotech Local Innovation System of Minas Gerais: University, Local Firms and Transnational Corporations. In *Biotechnology and Innovations Systems: The Role of Public Policy*, ed. B. Göransson and C.M. Pålsson, 13–57. Cheltenham: Edward Elgar.

Cassiolato, J.E., G. Zucoloto, R. Molagres, and F. Stalliveiri. 2014. Transnational Corporations and the Brazilian National System of Innovation. In *Transnational Corporations and Local Innovation*, ed. J.E. Cassiolato, G. Zucoloto, D. Abrol, and L. Xielin, 68–132. Abingdon: Routledge.

Cederlöf, G., and K. Sivaramakrishnan. 2006. *Ecological Nationalisms: Nature, Livelihoods, and Identities in South Asia*. Seattle, WA: University of Washington Press.

Cerny, P.G. 1997. Paradoxes of the Competition State: The Dynamics of Political Globalization. *Government and Opposition* 32: 251–274.

Cesarino, L., and N. Luna. 2011. The Embryo Research Debate in Brazil: From the National Congress to the Federal Supreme Court. *Social Studies of Science* 41 (2): 227–250.

Chamas, C.I., S. Paulino de Carvalho, and S. Salles-Filho. 2007. Current Issues of IP Management in Health and Agriculture in Brazil. In *Intellectual Property Management in Health and Agricultural Innovation: A Handbook of Best Practices*, ed. A. Krattiger, R.T. Mahoney, L. Nelsen, et al., 1563–1575. Oxford: MIHR.

Chary, T.S., and R.K. Mishra. 2008. *Venture Capital Financing for Biotechnology*. New Delhi: Concept Publishing.

Chatterjee, Partha. 1993. *The Nation and Its Fragments*. Princeton, NJ: Princeton University Press.

Chen, Z., H. Wang, Z. Wen, and Y. Wang. 2007. Life Sciences and Biotechnology in China. *Philosophical Transactions of the Royal Society of London* 362 (1482): 947–957.

Cheng, G., C. Wang, and D. Xue. 2012. Attitude to the Traditional Knowledge of International Governmental Organizations and Suggested Strategies of China. *Biodiversity Science* 20 (4): 505–511.

Chesnais, F. 1986. Science, Technology and Competitiveness. *STI Review* 1 (Autumn): 85–129.

———. 1991. Technological Competitiveness Considered as a Form of Structural Competitiveness. In *Technology and National Competitiveness: Oligopoly, Technological Innovation, and International Competition*, ed. J. Niosi, 142–176. Montreal: McGill-Queen's University Press.

———. 2004. Globalization Against Development. *International Socialism Journal* 102: 26–44.

Chibber, V. 2005. Reviving the Developmental State? The Myth of the "National Bourgeoisie". *Socialist Register* 41: 226–246.

———. 2012. Organized Interests, Development Strategies, and Social Policies. In *Growth, Inequality, and Social Policy in India*, ed. R. Nagaraj, 168–192. London: Palgrave Macmillan.

Chmielewska, D., and D. Souza. 2011. The Food Security Context in Brazil: Country Profile. International Policy Centre for Inclusive Growth (UNDP). Available at http://www.ipc-undp.org/pub/IPCCountryStudy22.pdf. Accessed 6 April 2017.

Chopra, R. 2011. Calorie Intake of the Rural Poor Has Drastically Fallen, Says Planning Commission Report. *India Today*, October 22. Available at http://indiatoday.intoday.in/story/poverty-malnutrition-india-hunger-index/1/157110.html. Accessed 6 April 2017.

Cimoli, M., B. Coriat, and A. Primi. 2008. *Intellectual Property and Industrial Development: A Critical Assessment*. Initiative for Policy Dialogue Working Paper Series. New York: Columbia University.

Clark, G.L. 2000. *Pension Fund Capitalism*. Oxford: Oxford University Press.

CMR (Centre for Medicines Research). 2010. *The CMR International Pharmaceutical R&D Factbook*. Chicago: Thomson Reuters.

Conklin, B.A. 2002. Shamans Versus Pirates in the Amazonian Treasure Chest. *American Anthropologist* 104 (4): 1050–1061.

Cooke, P. 2001. New Economy Innovation Systems: Biotechnology in Europe and the USA. *Industry and Innovation* 8 (3): 267–289.

Cooper, M. 2008. *Life as Surplus: Biotechnology and Capitalism in the Neoliberal Era*. Seattle: University of Washington Press.

Coriat, B., and O. Weinstein. 2009. *Intellectual Property Right Regimes, Firms and the Commodification of Knowledge*. Comparative Research in Law & Political Economy, Research Paper No. 17/2009. Available at http://digitalcommons.osgoode.yorku.ca/clpe/130. Accessed 20 May 2016.

Coriat, B., F. Orsi, and O. Weinstein. 2003. Does Biotech Reflect a New Science-Based Innovation Regime? *Industry and Innovation* 10 (3): 231–253.

Coriat, P., and G. Schméder. 2006. Post-Fordism in a More Globalized Capitalism. In *The Hardship of Nations: Exploring the Paths of Modern Capitalism*, ed. P. Coriat, P. Petit, and G. Schméder, 311–340. Cheltenham: Edward Elgar.

Correa, M., P. Villardi, and P. Villela 2014. Pharmaceuticals and Public Health in Brazil: Copying Essential Drugs, Knowledge Acquisition and Innovation Projects in Public-Private Industrial Networks. Unpublished draft, in file with author.

Coutinho, L.G. 2003. Macroeconomic Regimes and Business Strategies: An Alternative Industrial Policy for Brazil in the Wake of the 21st Century. In *System of Innovation and Development: Evidence from Brazil*, ed. M.L. Maciel. Cheltenham: Edward Elgar.

Coutinho, M., R.L.M. Silva, and M.A. Marin. 2001. Biodiversity and Public Policy Issues in Development: Brazil as a Case Study. Paper Presented at the International Society for History, Philosophy and Social Studies of Biology, Annual Meeting, Hamden, USA.

Crotty, J.R. 1990. Owner-Manager Conflict and Financial Theories of Investment Instability: A Critical Assessment of Keynes Tobin and Minsky. *Journal of Post-Keynesian Economics* 12 (4): 519–542.

Crosby, A. 1972. *The Columbian Exchange: Biological and Cultural Consequences of 1492*. Westport, CT: Greenwood Press.

Crow, J.F. 1998. 90 Years Ago: The Beginning of Hybrid Maize. *Genetics* 148 (March): 923–928.

Cumings, B. 1999. Webs with No Spiders, Spiders with No Webs: The Genealogy of the Developmental State. In *The Developmental State*, ed. M. Woo-Cumings. Ithaca, NY: Cornell University Press.

Cunha, J. 2015. Farmacêuticas Brasileiras Compram Empresas No Exterior por Inovação. *Folha de S.Paolo*. Available at http://www1.folha.uol.com.br/mercado/2015/06/1644256-farmaceuticas-brasileiras-compram-empresas-no-exterior.shtml. Accessed 20 September 2016.

Cunha, M.C., and M. Almeida. 2000. "Indigenous People, Traditional People and Conservation in the Amazon," in Brazil: Burden of the Past, Promise of the Future. *Journal of the American Academy of Sciences* 129 (2): 315–338.

Daniel, G.R. 2014. Machado de Assis and Brazilian Literary Independence: Toward a Postcolonial National Aesthetic. *Machado de Assis Em Linha* 7 (14): 163–182.

Daly, H.E., and J.B. Cobb. 1989. *For the Common Good*. Uckfield: Beacon Press.

Danzon, P., and E. Keuffel. 2005. *Regulation of the Pharmaceutical Industry*. Paper Presented at the National Bureau of Economic Research Conference on Regulation, 9–10 September 2005, Cambridge, MA.

Danzon, P., A. Epstein, et al. 2003. *Mergers and Acquisitions in the Pharmaceutical and Biotech Industries*. Organizational Economics of Health Care Conference 2003. Available at http://ssrn.com/abstract=468301. Accessed 30 May 2016.

Das, A.N. 1994. Brazil and India: Comparing Notes on Liberalisation. *Economic and Political Weekly* 29 (8): 416–419.

Dasgupta, Bilal. 1999. Patent Lies and Latent Danger. *Economics and Political Weekly* 17: 976–993.

Destler, I.M. 1992. *American Trade Politics*. New York: Institute for International Economics and the Twentieth Century Fund.

Dinca, R. 2005. The "Bermuda Triangle" of Pharmaceutical Law: Is Data Protection a Lost Ship? *The Journal of World Intellectual Property* 8 (4): 517–563.

Drahos, P. 1996. *A Philosophy of Intellectual Property*. Aldershot: Ashgate.

———. 2002. Developing Countries and International Intellectual Property Standard-Setting. *The Journal of World Intellectual Property* 5: 765–789.

Drahos, P., and J. Braithwaite. 2002. *Informational Feudalism: Who Owns the Knowledge Economy?* London: Earthscan.

Drahos, Peter. 2010. *The Global Governance of Knowledge*. Cambridge: Cambridge University.

Drèze, J., and A. Sen. 1989. *Hunger and Public Action*. Oxford: Oxford University.

Dutfield, G. 2004. *Intellectual Property, Biogenetic Resources and Traditional Knowledge*. Abingdon: Earthscan.

———. 2011. The Role of the International Union for the Protection of New Varieties of Plants (UPOV). Global Economic Issue Publications, Intellectual Property Issue Paper Number 9, Quakers United Nations Office, Geneva. Available at http://quno.org/sites/default/files/resources/UPOV%2Bstudy%2Bby%2BQUNO_English.pdf. Accessed 6 June 2016.

———. 2003. *Intellectual Property and the Life Sciences Industries: A Twentieth Century History*. Aldershot: Dartmouth Publishing.

Dutz, Mark, ed. 2007. *Unleashing India's Innovation: Toward Sustainable and Inclusive Growth*. Washington, DC: World Bank.

EC (European Commission). 1996. "Biotechnology: State of Play in Europe" Press Release 11th January 1996. Available at http://europa.eu/rapid/press-release_MEMO-96-2_en.htm. Accessed 5 April 2017.

EFPIA. 2004. *Annual Report Brussels, European Federation of Pharmaceutical Industries and Associations*. European Federation of Pharmaceutical Industries and Associations, Brussels, Belgium (in file with author).

Eimer, T.R., S. Lütz, and V. Verena Schüren. 2016. Varieties of Localization: International Norms and the Commodification of Knowledge in India and Brazil. *Review of International Political Economy* 23 (3): 450–479.

Emmert, F. 1990. Intellectual Property in the Uruguay Round – Negotiating Strategies of the Western Industrialized Countries. *Michigan Journal of International Law* 11: 1317–1399.

Evans, P. 1979. *Dependent Development: the Alliance of Multinational, State and Local Capital in Brazil*. Princeton, NJ: Princeton University Press.

———. 2008. *In Search of the 21st Century Developmental State*. Working Paper no. 8, Centre for Global Political Economy, University of Sussex, UK.

Feijo, C.A., and M.T. Lamonica. 2010. The Importance of the Manufacturing Sector for Brazilian Economic Development. *CEPAL Review* 102: 7–26.

FHC (Fernando Henrique Cardoso). 2000. Discurso do presidente da República na cerimônia de apresentação do "Avança Brasil": Plano Plurianual 2000/2003 e do Orçamento da União para o ano 2000, Palácio do Planalto, Brasília. Available at http://www1.folha.uol.com.br/fsp/brasil/fc01099911.htm. Accessed 22 September 2016.

———. 2002. O Brasil a caminho da sociedade do conhecimento. In *O Brasil e a Economia do Conhecimento*, ed. J.P. Velloso. Rio de Janeiro: J. Olympio.

Filoche, G. 2012. Biodiversity Fetishism and Biotechnology Promises in Brazil: From Policy Contradictions to Legal Adjustments. *The Journal of World Intellectual Property* 15 (2): 133–154.

Filomeno, F. 2013. State Capacity and IP Regimes: Lessons from South American Soybean Agriculture. *Technology in Society* 35: 139–152.

———. 2014a. *Monsanto and Intellectual Property in South America*. Basingstoke: Palgrave Macmillan.

———. 2014b. Patterns of Rule-Making and Intellectual Property Regimes. *Comparative Politics* 46 (4): 439–458.

Fisk, C. 2003. Authors at Work: The Origins of the Work-for-Hire Doctrine. *Yale Journal of Law and Humanities* 15: 1–69.

Fleury, S. 2014. Building Democracy in an Emerging Society: Challenges of the Welfare State in Brazil. In *Brazil Emerging: Inequality and Emancipation*, ed. J.N. Pieterse and A. Cardoso, 11–31. Abingdon: Routledge.

Flynn, M. 2011. Corporate Power and State Resistance: Brazil's Use of TRIPS Flexibilities. In *Intellectual Property, Pharmaceuticals and Public Health: Access to Drugs in Developing Countries*, ed. K. Shadlen, 149–177. Cheltenham: Edward Elgar Publishing.

———. 2013. Origins and Limitations of State-based Advocacy: Brazil's AIDS Treatment Program and Global Power Dynamics. *Politics and Society* 41 (1): 3–28.

Fortun, M. 2012. Genomics Scandals and Other Volatilities of Promising. In *Lively Capital: Biotechnologies, Ethics and Governance in Global Markets*, ed. K. Rajan, 329–353. Durham, NC: Duke University Press.

Fowler, C. 1994. *Unnatural Selection: Technology, Politics and Plant Evolution*. Verdon, Switzerland: Gordon & Breach.

Furlan, L.F., A. Álvares, L.C.G. Pinto, and S. Rezende. 2006. Futuro Certo e Promissor da Biotecnologia. *Folha de São Paulo*, June 23.

Furtado, C. 1992. *Brasil: A Construção Interrompida*. São Paolo: Paz e Terra.

———. 1997. *Obra Autobiográfica de Celso Furtado*. São Paolo: Paz e Terra.

Gadelha, C.A.G., L.S. Costa, J.M.S.V. Maldonado, and M.A. Vargas. 2013. The Healthcare Economic-Industrial Complex: Concepts and General Characteristics. *Health* 5 (10): 1607–1621.

Galvão, J. 2000. *A AIDS No Brasil: A Agenda de Construção de Uma Epidemia*. São Paulo: Editora 34.

Garfield, S. 2013. *In Search of the Amazon: Brazil, the United States and the Nature of a Region*. London: Duke University Press.

Gereffi, G. 1983. *The Pharmaceutical Industry and Dependency in the Third World*. Princeton, NJ: Princeton University Press.

Gerschenkron, A. 1962. *Economic Backwardness in Historical Perspective*. Cambridge: Belknap Press.

GFB (Grupo Farma Brasil). 2014. *The Rise of a National Biopharmaceutical Industry in Brazil: Consolidation, Trends and Challenges*. Available at https://c.ymcdn.com/sites/casss.site-ym.com/resource/resmgr/CMC_LATAM_Speaker_Slides/2014_CMCLAT_ArcuriReginaldo.pdf. Accessed 21 September 2016.

Glenna, L.L., and D.R. Cahoy. 2009. Agribusiness Concentration, Intellectual Property and the Prospects for Rural Economic Benefits from the Emerging Biofuel Economy. *Southern Rural Sociology* 24 (2): 111–129.

Godin, B. 2006. The Knowledge-Based Economy: Conceptual Framework or Buzzword? *Journal of Technology Transfer* 31: 17–30.

GoI. 2003. *Science and Technology Policy 2003*. Available at http://www.csir.res.in/external/heads/aboutcsir/Policy2003.htm. Accessed 11 July 2017.

———. 2007. *National Biotechnology Development Strategy*. New Delhi: Department of Biotechnology, Ministry of Science and Technology.

———. 2012. *Invitation of Views on the Draft National IPR Strategy*. New Delhi: Department of Industrial Policy and Promotion.

———. 2013. *Science*. New Delhi: Technology and Innovation Policy, Ministry of Science and Technology.

———. 2014. Agricultural Census: All Indian Report on Number and Area of Operational Holdings, 2010–11, Ministry of Agriculture. Available at http://agcensus.nic.in/document/agcensus2010/CompleteReport.pdf. Accessed 15 July 2016.

———. 2015. *Draft National Health Policy*. Ministry of Health & Family Welfare. Available at http://www.mohfw.nic.in/showfile.php?lid=3014. Accessed 20 February 2017.

Gompers, P. 1994. The Rise and Fall of Venture Capital. *Business and Economic History* 23 (2): 1–26.

Gontijo, C. 2005. *Changing the Patent System: From the Paris Convention to the TRIPS Agreement*. Global Issue Paper No. 26, The Heinrich Böll Foundation, Berlin.

Görg, C., and U. Brand. 2000. Global Environmental Politics and Competition Between Nation-States: On the Regulation of Biological Diversity. *Review of International Political Economy* 7 (3): 371–398.

Gosain, R. 2007. In Good Health? Compulsory Licenses. *Patent World* 194: 21–30.

Gouvea, R., and S. Kassicieh. 2005. Using Resources in R&D Policy Planning: Brazil, the Amazon and Biotechnology. *Technological Forecasting & Social Change* 72: 535–547.

Goven, J., and V. Pavone. 2015. 'The Bioeconomy as Political Project: A Polanyian Analysis' *Science. Technology & Human Values* 40 (3): 302–337.

Greene, W. 2007. *The Emergence of India's Pharmaceutical Industry and Implications for the U.S. Generic Drug Market*. The US International Trade Commission, Working Paper No. 2007-05-A. Available at https://www.usitc.gov/publications/332/EC200705A.pdf. Accessed 10 July 2016.

Gross, R. 2014. *National Study on ABS Implementation in Brazil*. The ABS Capacity Development Initiative and the Government of Brazil. Available at http://www.abs-initiative.info/fileadmin/media/Knowledge_Center/Pulications/ABS_Dialogue_042014/National_study_on_ABS_implementation_in_Brazil_20140716.pdf. Accessed 25 January 2017.

Gupta, A.S. 2010. B K Keayla: A Personal Reminiscence. *Economic and Political Weekly* 45 (51): 25–26.

Harris, J.R. 1998. *Industrial Espionage and Technology Transfer: Britain and France in the Eighteenth Century*. Aldershot: Ashgate.

Harvey, M. 2002. Competition as Instituted Economic Process. In *Market Relations and the Competitive Process*, ed. S. Metcalfe and A. Warde. Manchester: Manchester University Press.

Harvey, M., and A. McMeekin. 2005. Brazilian Genomics and Bioinformatics: Instituting New Innovation Pathways in a Global Context. *Economy and Society* 34 (4): 634–658.

Hay, C. 2004. Re-Stating Politics, Re-Politicising the State. In *Restating the State?* ed. A. Gamble and T. Wright, 38–50. Oxford: Blackwell.

Heller, M., and R. Eisenberg. 1998. Can Patents Deter Innovation? The Anticommons in Biomedical Research. *Science* 280 (5364): 698–701.

Henderson, E. 1997. TRIPs and the Third World: The Example of Pharmaceutical Patents in India. *European Intellectual Property Review* 11: 651–663.

Herring, R.J. 1999. Embedded Particularism: India's Failed Developmental State. In *The Developmental State*, ed. M. Woo-Cummings, 306–344. Ithaca, NY: Cornell University Press.

Hester, R.J. 2016. Biology as Opportunity: Hybrid Rule from a Molecular View. In *Hybrid Rule and State Formation: Public-Private Power in the 21st Century*, ed. S.L. Hurt and R.D. Lipschutz, 175–192. London: Routledge.

Hobson, J. 1997. *The Wealth of States: A Comparative Sociology of International Economic and Political Change*. Cambridge: Cambridge University Press.

Hopkins, M.M., P.A. Martin, P. Nightingale, et al. 2007. The Myth of the Biotech Revolution: An Assessment of Technological, Clinical and Organisational Change. *Research Policy* 36 (4): 566–589.

Howard, P.H. 2015. Intellectual Property and Consolidation in the Seed Industry. *Crop Science* 55: 1–7.

Hsü, I.C.Y. 2000. *The Rise of Modern China*. Oxford: Oxford University Press.

Huang, J., S. Rozelle, C. Pray, and Q. Wang. 2002. Plant Biotechnology in China. *Science* 295: 674–676.

Hughes, J. 1997. The Philosophy of Intellectual Property. In *Intellectual Property: Moral, Legal and International Dilemmas*, ed. A. Moore. Oxford: Rowman & Littlefield Publishers.

Hurrell, A. 2005. The United States and Brazil: Comparative Reflections. In *The United States and Brazil: A Long Road of Unmet Expectations*, ed. M. Hirst, 73–106. London: Routledge.

———. 2005. The United States and Brazil: Some Comparative Reflections. In *The United States and Brazil: A Long Road of Unmet Expectations*, ed. M. Hirst, 75–90. New York: Routledge.

Hurt, S.L. 2016. What's at Stake in the Privatization Debate? Enclosing the Public Domain Through Hybrid Rule. In *Hybrid Rule and State Formation: Public-Private Power in the 21st Century*, ed. S.L. Hurt and R.D. Lipschultz, 101–122. London: Routledge.

IDMA. 2009. Patents. Available at http://www.idma-assn.org/patents.html. Accessed 5 July 2016.

ILO. 2012. Statistical Update on Employment in the Informal Economy, ILO – Department of Statistics. Available at http://laborsta.ilo.org/applv8/data/INFORMAL_ECONOMY/2012-06-Statistical%20update%20-%20v2.pdf. Accessed 6 April 2017.

IMF. 2010. World Economic Outlook. Available at https://www.imf.org/external/pubs/ft/weo/2010/01/. Accessed 10 March 2016.

Islam, N. 2010. Indigenous Medicine as Commodity. *Current Sociology* 58 (5): 777–798.

Iyer, K.V.R. 2002. The Doha-Delhi Odyssey. *Frontline* 19 (6): 16–29.

Izique, C. 2002. Actions Against Biopiracy. *Revista Pesquisa FAPESP*, June Issue. Available at http://revistapesquisa.fapesp.br/en/2002/06/01/actions-against-biopiracy/. Accessed 22 September 2016.

Jakobson, L. 2007. China Aims High in Science and Technology: An Overview of Challenges Ahead. In *Innovation with Chinese Characteristics: High Tech Research in China*, ed. L. Jakobson, 1–36. Basingstoke: Palgrave Macmillan.

James, C. 2015. 20th Anniversary (1996 to 2015) of the Global Commercialization of Biotech Crops and Biotech Crop Highlights in 2015. ISAAA Brief No. 51. Available at http://isaaa.org/resources/publications/briefs/51/executivesummary/default.asp. Accessed 6 June 2016.

Jasanoff, S. 2005. *Designs on Nature: Science and Democracy in Europe and the United States*. Princeton, NJ: Princeton University Press.

Jenkins, R. 1999. *Democratic Politics and Economic Reform in India*. Cambridge: Cambridge University Press.

———. 2003. Neo-Liberalism Encounters India's Indigenous Political Traditions. *Economy and Society* 32 (4): 584–610.

Jepson, W.E. 2002. Globalization and Brazilian Biosafety: The Politics of Scale Over Biotechnology Governance. *Political Geography* 21: 905–925.

Jepson, W.E., C. Brannstrom, and R.S. de Souza. 2008. Brazilian Biotechnology Governance: Consensus and Conflict Over Genetically Modified Crops. In *Food For the Few: Neoliberal Globalism and Biotechnology in Latin America*, ed. G. Otero, 217–242. Austin: University of Texas Press.

Jessop, B. 1993. Towards a Schumpeterian Workfare State? Preliminary Remarks on Post-Fordist Political Economy. *Studies in Political Economy* 40: 7–40.

———. 1995. The Regulation Approach, Governance and Post-Fordism: Alternative Perspectives on Economic and Political Change? *Economy and Society* 24 (3): 307–333.

———. 1997. Capitalism and Its Future: Remarks on Regulation, Government and Governance. *Review of International Political Economy* 4 (3): 561–581.

———. 2000. The State and Contradictions of the Knowledge-Driven Economy. In *Knowledge, Space, Economy*, ed. J.R. Bryson, P.W. Daniels, N.D. Henry, and J. Pollard, 63–78. London: Routledge.

———. 2001. What Follows Fordism? On the Periodization of Capitalism and Its Regulation. In *Phases of Capitalist Development: Booms, Crises and Globalization*, ed. R. Albritton, M. Itoh, R. Westra, and A. Zuege, 282–299. Basingstoke: Palgrave.

———. 2002. *The Future of the Capitalist State*. Cambridge: Polity.

———. 2005. Cultural Political Economy, the Knowledge Based Economy and the State. In *The Technological Economy*, ed. A. Barry and D. Slater, 142–164. London: Routledge.

———. 2007. Knowledge as Fictitious Commodity. In *Reading Polanyi for the 21st Century*, ed. A. Bugra. Basingstoke: Palgrave Macmillan.

———. 2010. The Knowledge Economy as a State Project. In *The Nation-State in Transformation*, ed. M. Böss, H. Bang, and J. Campbell, 110–129. Santa Barbara: Aarhus University Press.

Johnson, C. 1982. *MITI and the Japanese Miracle: The Growth of Industrial Policy: 1925–1975*. Stanford, CA: Stanford University Press.

———. 1999. The Developmental State: Odyssey of a Concept. In *The Developmental State*, ed. M. Woo-Cumings, 32–60. New York: Cornell University Press.

Joseph, K.J., and D. Abrol. 2009. Science, Technology and Innovation Policies in India: Achievements and Limits. In *BRICS and Development Alternatives: Innovation Systems and Policies*, ed. J.E. Cassiolato and V. Vitorino, 101–131. London: Anthem Press.

Kapczynski, A. 2009. Harmonization and Its Discontents: A Case Study of TRIPS Implementation in India's Pharmaceutical Sector. *California Law Review* 97: 1570–1649.

Kaplan, W., and R. Laing. 2005. *Local Production of Pharmaceuticals: Industrial Policy and Access to Medicines*. Available at http://siteresources.worldbank.org/HEALTHNUTRITIONANDPOPULATION/Resources/281627-1095698140167/KaplanLocalProductionFinal.pdf. Accessed 10 November 2015.

Kavani, S. 1999. R A Mashelkar-Catalyst for Change. *Business India*, June 28–July 11.

Keayla, B.K. 2005. Amended Patents Act: A Critique. *Combat Law* 4 (2).

Keck, M. 2002. Amazônia in Environmental Politics. In *Environment and Security in the Amazon Basin*, ed. J.S. Tulchin and H.A. Golding, 31–52. Washington, DC: Woodrow Wilson International Center for Scholars.

Keeley, J. 2005. Interrogating China's Biotechnology Revolution: Contesting Dominant Science Policy Cultures. In *Science and Citizens*, ed. I. Scoones, M. Leach, and B. Wynne, 155–166. London: Zed Books.

Kenway, J., E. Bullen, J. Fahey, and S. Robb, eds. 2006. *Haunting the Knowledge Economy*. London: Routledge.

Kesič, D. 2011. Strategic Development Trends in the World Pharmaceutical Industry. *Managing Global Transitions* 9 (3): 207–223.

Khilnani, S. 1997. *The Idea of India*. London: Hamish Hamilton.

Kloppenburg, J. 2004. *First the Seed. The Political Economy of Plant Biotechnology, 1492–2000*. 2nd ed. Madison: Wisconsin University Press.

———. 2010. 'Impending Dispossession, Enabling Repossession: Biological Open Source and the Recovery of Seed Sovereignty. *Journal of Agrarian Change* 10 (3): 367–388.

Koeller, P., and J.E. Cassiolato. 2009. Achievements and Shortcomings of Brazil's Innovation Policies. In *BRICS and Development Alternatives: Innovation Systems and Policies*, ed. J.E. Cassiolato and V. Vitorino, 35–71. London: Anthem Press.

Koeller, P., and J.L. Gordon. 2013. Brazil. In *The State and the National System of Innovation: A Comparative Analysis of the BRICS Economies*, ed. M. Scerri and H.M.M. Lasters, 23–79. London: Routledge.

Kohli, A. 2004. *State-Directed Development: Political Power and Industrialization in the Global Periphery*. Cambridge: Cambridge University Press.

———. 2006. Politics of Economic Growth in India, Part I & II. *Economic and Political Weekly* 6: 1361–1370.

———. 2012. State and Redistributive Development in India. In *Economic Growth, Inequality and Social Development in India: Is Inclusive Growth Possible?* ed. R. Nagaraj, 194–226. Basingstoke: Palgrave Macmillan.

Kohli, K., M. Fareedi, and S. Shalini Bhutani. 2009. *Six Years of the Biological Diversity Act in India*. Kalpavriksh and GRAIN, Delhi/Pune. Available at https://www.grain.org/article/entries/4172-6-years-of-the-biological-diversity-act-in-india. Accessed 15 July 2016.

Kolk, A. 1996. *Forests in International Environmental Politics: International Organisations, NGOs, and the Brazilian Amazon*. Utrecht: International Books.

Krishnaswamy, S. 2011. Access to Knowledge and Traditional Knowledge Protection. In *Access to Knowledge in India*, ed. Ramesh Subramanian and Lea Shaver. Gordonsville, VA: Bloomsbury Publishing.

Krishna, V.V. 1997. A Portrait of the Scientific Community in India: Historical Growth and Contemporary Problems. In *Scientific Communities in the*

Developing World, ed. J. Gaillard, V.V. Krishna, and R. Waast, 236–280. New Delhi: Sage Publications.

———. 2013. India. In *BRICS National Systems of Innovation: The Role of the State*, ed. M. Scerri and H.M.M. Lastres, 138–187. New Delhi: Routledge.

Kumar, N., U. Quach, et al. 2004. Indian Biotechnology – Rapidly Evolving and Industry Led. *Nature Biotechnology* 22 (12): 31–36.

Lamounier, B. 1989. Brazil; Inequality Against Democracy. In *Democracy in Developing Countries*, ed. L. Diamond et al., vol. 4. Boulder, CO: Lynne Rienner Publisher.

Landes, W.M., and R.A. Posner. 2004. *The Political Economy of Intellectual Property Law*. Washington, DC: AEI Press.

Lastres, H.M.M., J.E. Cassiolato, and M.L. Maciel. 2003. Systems of Innovation for Development in the Knowledge Era. In *Systems of Innovation and Development*, ed. J.E. Cassiolato, H.M.M. Lastres, and M.L. Maciel, 1–32. Cheltenham: Edward Elgar Publishing.

Leamon, A., and J. Lerner. 2012. Creating a Venture Ecosystem in Brazil: FINEP's INOVAR Project. Working Paper 12-099, May 8, 2012, Harvard Business School.

Leiva, F.I. 2008. *Latin American Neostructuralism: The Contradictions of Post-Neoliberal Development*. Minneapolis: University of Minnesota Press.

Leubolt, B. 2014. *Social Policies and Redistribution in Brazil*. Working Paper No. 26. Geneva: International Labour Office.

Lipietz, A. 2001. The Fortunes and Misfortunes of Post-Fordism. In *Phases of Capitalist Development: Booms, Crises and Globalizations*, ed. R.M. Albritton, R. Westra, and A. Zuege, 17–36. Basingstoke: Palgrave.

Lula (Luiz Inácio da Silva). 2007. Discurso do Presidente da República, Luiz Inácio Lula da Silva, na cerimônia de lançamento da Política de Biotecnologia, Palácio do Planalto, Brasília, 08 de fevereiro de 2007. Available at http://www.biblioteca.presidencia.gov.br/presidencia/ex-presidentes/luiz-inacio-lula-da-silva/discursos/

Macdonald, S. 2002. Exploring the Hidden Costs of Patents. In *Global Intellectual Property Rights: Knowledge Access and Development*, ed. P. Drahos and R. Mayne. Basingstoke: Palgrave Macmillan.

Mahtaney, P. 2007. *India, China and Globalization*. Basingstoke: Palgrave Macmillan.

Marques, M.B. 1999. Gestão, planejamento e avaliação de políticas de ciência e tecnologia: hora de rever? *Ciência & Saúde Coletiva* 4 (2): 383–392.

Matthews, D. 2002. *Globalising Intellectual Property Rights: The TRIPs Agreement*. London: Routledge.

———. 2011. *Intellectual Property, Human Rights and Development: The Role of NGOs and Social Movements*. Cheltenham: Edward Elgar.

May, C. 2000. *A Global Political Economy of Intellectual Property Rights: The New Enclosures?* London: Routledge.

May, C., and S.K. Sell. 2006. *Intellectual Property Rights: A Critical History.* London: Lynne Reinner Publishers.

Mazzoleni R., and L.M.C. Póvoa. 2009. Accumulation of Technological Capabilities and Economic Development. Available at http://ideas.repec.org/f/pma824.html. Accessed 4 October 2016.

Mazzucato, M. 2015. *The Entrepreneurial State: Debunking Public vs. Private Sector Myths.* London: Anthem Press.

McAfee, K. 1999. Selling Nature to Save It? Biodiversity and Green Developmentalism. *Environment and Planning D: Society and Space* 17: 133–154.

Merges, R.P. 2000. One Hundred Years of Solicitude: Intellectual Property Law, 1900–2000. *California Law Review* 88 (6): 2178–2240.

Mooney, P.R. 1983. *The Law of the Seed: Another Development and Plant Genetic Resources.* Uppsala: Dag Hammarskjöld Foundation.

Morris-Suzuki, T. 1986. Capitalism in the Computer Age. *New Left Review* 160 (November–December): 81–91.

Mouhoud, E.M. 2006. The Economic Geography of Post Fordism: On Knowledge and Polarization. In *The Hardship of Nations: Exploring the Paths of Modern Capitalism*, ed. B. Coriat, P. Petit, and G. Schméder, 293–307. Cheltenham: Edward Elgar.

Mueller, B., and C. Mueller. 2014. *The Economics of the Brazilian Agricultural Development.* Working Paper, International Research Initiative on Brazil and Africa (IRIBA), University of Manchester, UK.

Mueller, J. 2007. The Tiger Awakens: the Tumultuous Transformation of India's Patent System and the Rise of Indian Pharmaceutical Innovation. *University of Pittsburgh Law Review* 68: 491–641.

———. 2008. *Biotechnology Patenting in India: Will Bio-Generics Lead a Sunrise Industry to Bio-Innovation?* Legal Studies Research Paper Series, University of Pittsburgh School of Law, Working Paper No. 2008-2.

Mukherji, R. 2012. Ideas, Interests and the Tipping Point: Economic Change in India. *Review of International Political Economy* 20 (2): 363–389.

Murphy, C.N. 1983. What the Third World Wants: An Interpretation of the Development and Meaning of the New International Economic Order Ideology. *International Studies Quarterly* 27 (1): 55–76.

Murphy, S., D. Burch, and J. Clapp. 2012. *Cereal Secrets: The world's Largest Grain Traders and Global Agriculture.* Oxfam Research Report. Available at https://www.oxfam.org/sites/www.oxfam.org/files/rr-cereal-secrets-grain-traders-agriculture-30082012-en.pdf. Accessed 6 June 2016.

Muzaka, V. 2011. *The Politics of Intellectual Property Rights and Access to Medicines.* Basingstoke: Palgrave Macmillan.

———. 2013. Contradictions, Frames and Reproductions: The Emergence of the WIPO Development Agenda. *Review of International Political Economy* 20 (1): 215–239.

———. 2015. The State as Facilitator and Legitimator of 'New' Capital Accumulation: The Case of Patent Reform in India. *Journal of International Relations and Development* iFirst: 1–24.

———. 2017. Lessons from Brazil: On the Difficulties of Building a Universal Healthcare System. *The Journal of Global Health* 7 (1): 010303.

Nastari, P.M. 1983. *The Role of Sugar Cane in Brazil's History and Economy*. Retrospective Theses and Dissertations, n. 8948, Digital Repository of the Iowa State University.

Nayar, B.R. 1979. A World Role: The Dialectics of Purpose and Power. In *India: A Rising Middle Power*, ed. John W. Mellor. Boulder, CO: Westview Press.

———. 2000. The Limits of Economic Nationalism in India: Economic Reforms under the BJP-Led Government. *Asian Survey* 40 (5): 792–815.

Nehring, R. 2016. *Yields of Dreams: Marching West and the Politics of Scientific Knowledge in the Brazilian Agricultural Research Corporation (Embrapa)*. Paper No. 35 Presented at Global Governance Climate Justice and Agrarian Justice International Colloquium, 4–5 February 2016 (The Hague: The Netherlands).

Newell, P. 2003. *Biotech Firms, Biotech Politics: Negotiating GMOs in India*. Institute of Development Studies, Working Paper 201, Brighton, UK.

———. 2008. Trade and Biotechnology in Latin America: Democratization, Contestation and the Politics of Mobilization. *Journal of Agrarian Change* 8 (2/3): 245–376.

Ni, X. 2015. China's Research & Development Spend. *Nature* 520: 8–9.

Niosi, J., and P. Faucher. 1991. The State and International Trade: Technology and Competitiveness. In *Technology and National Competitiveness: Oligopoly, Technological Innovation, and International Competition*, ed. J. Niosi, 119–141. Montreal: McGill-Queen's University Press.

Nunn, A. 2008. *The Politics and History of AIDS Treatment in Brazil*. New York: Springer.

Octaviani, A. 2010. Biotechnology in Brazil: Promoting Open Innovation. In *Access to Knowledge in Brazil*, ed. L. Sheaver, 79–102. London: Bloomsbury Academic.

OECD. 2001. *Using Knowledge for Development: The Brazilian Experience*. Paris: OECD Publishing.

———. 2006. *Scoping Document: The Bioeconomy to 2030 – Designing a Policy Agenda*. Paris: OECD.

Okediji, R. 2003. The International Relations of Intellectual Property: Narratives of Developing Country Participation in the Global Intellectual Property System. *Singapore Journal of International and Comparative Law* 7: 315–385.

Orsenigo, L. 1989. *The Emergence of Biotechnology: Institutions and Markets in Industrial Innovation*. London: Pinter.

Ostry, S. 2000. Looking Back to Look Forward: The Multilateral Trading System After Fifty Years. In *From GATT to the WTO: The Multilateral Trading System in the New Millennium*, ed. WTO Secretariat, 97–112. The Hague: Kluwer Law International and the WTO.

Pagano, U., and A. Rossi. 2009. The Crash of Knowledge Economy. *Cambridge Journal of Economics* 33: 665–683.

Palma, J.G. 2012. *Brazil's Recent Growth*. UNCTAD and South Centre Discussion Paper No. 3 (Geneva: UNCTAD).

Pansera, C. 2016. Interview of Celso Pansera with Agência Brasil. Available at http://confap.org.br/news/brasil-vai-passar-de-pais-de-commodities-a-exportador-de-inovacao-diz-ministro/. Accessed 28 September 2016.

Park, C., and A. Jayadev. 2011. Access to Medicines in India: A Review of Recent Concerns. In *Access to Knowledge in India: New Research on Intellectual Property, Innovation and Development*, ed. L. Shaver and R. Subramanian, 78–108. London: Bloomsbury.

Patel, R. 2007. *Stuffed and Starved: Markets, Power and the Hidden Battle for the World Food System*. London: Portobello Books.

———. 2013. The Long Green Revolution. *The Journal of Peasant Studies* 40 (1): 1–63.

Paulani, L. 2010. Brazil in the Crisis of the Finance-led Regime of Accumulation. *Review of Radical Political Economics* 42 (3): 363–372.

Pedersen, J.D. 2008. *Globalization, Development and the State: The Performance of India and Brazil Since 1990*. Basingstoke: Palgrave Macmillan.

Peschard, K. 2012. Unexpected Discontent: Exploring New Developments in Brazil's Transgenics Controversy. *Canadian Journal of Development Studies* 33 (3): 326–337.

Petit, M., C. Fowler, et al. 2001. *Why Governments Can't Make Policy: The Case of Plant Genetic Resources in the International Arena*. Paper Prepared for the Global Forum on Agricultural Research. Available at http://agris.fao.org/agris-search/search.do?recordID=XF2016008984. Accessed 6 June 2016.

Petit, P. 1999. Structural Forms and Growth Regimes of the Post-Fordist Era. *Review of Social Economy* 57 (2): 220–243.

Philippou, S. 2005. Modernism and National Identity in Brazil, or How to Brew a Brazilian Stew. *National Identities* 7 (3): 245–264.

Pimentel, V.P., R.P. Gomes, T.L. Mitidieri, et al. 2014. *Inserção Internacional das Empresas Farmacêuticas: Motivações, Experiências e Propostas para o BNDES*. BNDES Report. Available at https://web.bndes.gov.br/bib/jspui/bitstream/1408/3107/1/BS%2040%20Inserção%20internacional%20das%20empresas%20farmacêuticas_P.pdf. Accessed 22 September 2016.

Pisano, G.P. 2006. *Science Business: The Promise, the Reality and the Future of Biotech*. Boston, MA: Harvard Business School Press.

PITCE. 2003. Diretrizes de Política Industrial, Tecnológica e de Comércio Exterior, 26 de Novembro de 2003. Available at http://www.inovacao.unicamp.br/politicact/diretrizes-pi-031212.pdf. Accessed 10 September 2016.

Pordié, L., and J.-P. GAudillière. 2014. The Reformulation Regime in Drug Discovery: Revisiting Polyherbals and Property Rights in the Ayurvedic Industry. *East Asian Science, Technology and Science* 8: 57–79.

Porter, T. 1999. Hegemony and the Private Governance of International Industries. In *Private Authority and International Affairs*, ed. A.C. Culter, V. Haufler, and T. Porter. Albany: State University of New York Press.

Portilho, D., and R. Gosain. 2007. The Pharmaceutical IP Environment in Brazil. *Managing Intellectual Property*. Available at http://www.managingip.com/Article/1450365/Brazil-Drugs-dispute-demands-action.html. Accessed 28 September 2016.

Prasad, A. 2009. Capitalizing Disease: Biopolitics of Drug Trials in India. *Theory, Culture, Society* 26 (5): 1–29.

Rada, N. 2013. Assessing Brazil's Cerrado Agricultural Miracle. *Food Policy* 38: 146–155.

Raghavan, C. 2003. Does India Have a Trade Policy? *The Hindu*, April 28.

Rajan, K.S. 2005. Subjects of Speculation: Emergent Life Sciences and Market Logics in the United States and India. *American Anthropologist* 107 (1): 19–30.

———. 2006. *Biocapital: The Constitution of Postgenomic Life*. Durham, NC: Duke University Press.

———. 2007. Biocapital as an Emergent Form of Life: Speculations on the Figure of the Experimental Subject. In *Biosocialities, Genetics and the Social Sciences: Making Biologies and Identities*, ed. S. Gibbon and C. Novas, 157–186. London: Routledge.

Rajan, K. 2012. The Capitalization of Life and the Liveliness of Capital. In *Lively Capital: Biotechnologies, Ethics, and Governance in Global Markets*, ed. K. Rajan, 1–43. London: Duke University Press.

Rajan, M.G. 1994. *India and the North-South Politics of Global Environmental Issues: The Cases of Ozone Depletion, Climate Change and Loss of Biodiversity*. Unpublished Ph.D. Thesis, Oxford University, Oxford.

Ramakrishna, T. 2002. *Development of IPR Regime in India with Reference to Agricultural Biotechnology*. Centre for Intellectual Property Rights, Research and Advocacy (CIPRA), National Law School of India University, Karnataka. Available at https://www.ids.ac.uk/files/Ramakrishna_IPR.pdf. Accessed 10 July 2016.

Ramanathan, U. 1996. Displacement and the Law. *Economic and Political Weekly*, June 15, pp. 1486–1491.

———. 2011. Land Acquisition, Eminent Domain and the 2011 Bill. *Economic and Political Weekly* 46 (44–45): 10–14.

Ramanna, A. 2002. Policy Implications of India's Patent Reforms: Patent Applications in the Post-1995 Era. *Economic and Political Weekly* 37 (21): 2065–2075.

———. 2005. Shifts in India's Policy on IP: The Role of Ideas, Coercion and Changing Interests. In *Death of Patents*, ed. P. Drahos, 150–174. London: Lawtext Publishing.

———. 2006a. *India's Policy on Genetically Modified Crops*. Asia Research Centre, Working Paper 15. London: LSE.

———. 2006b. *Farmers' Rights in India: A Case Study*. Background Study 4, The Farmers Rights Project. Lysaker, Norway: The Fridtjof Nansen Institute.

Ramirez, P., and A. Tylecote 1999. *Technological Change in the Pharmaceutical Industry: A Literature Review from the Point of View of Corporate Governance*. COPI Report Prepared for EC DG XII. Sheffield University, UK.

Randeria, S. 2003. Glocalization of Law. *Current Sociology* 52 (3/4): 305–328.

———. 2007. The State of Globalization: Legal Plurality, Overlapping Sovereignties and Ambiguous Alliances between Civil Society and the Cunning State in India. *Theory, Culture, Society* 24 (1): 1–33.

Rangnekar. 2006. No Pills for Poor People? *Economic and Political Weekly*, 4 February: 409–17.

Ranjan, P. 2009. Recent Developments in India's Plant Variety Protection, Seed Regulation and Linkages with UPOV's Proposed Membership. *The Journal of World Intellectual Property* 12 (3): 219–243.

Rao, C.M. 2004. Indian Seed System and Plant Variety Protection. *Economic and Political Weekly* 39 (8): 845–852.

Rasmussen, N. 2014. *Gene Jockeys: Life Science and the Rise of Biotech Enterprise*. Baltimore, MD: John Hopkins University Press.

Raustiala, K., and D. Victor. 2004. The Regime Complex for Plant Genetic Resources. *International Organization* 58 (2): 277–309.

Redwood, H. 1994. *New Horizons in India: The Consequences of Pharmaceutical Patent Protection*. Felixstowe: Oldwicks Press.

Reddy, K.S., V. Patel, P. Jha, et al. 2011. Towards Achievement of Universal Health Care in India by 2020: A Call to Action. *Lancet* 377: 760–768.

Reichman, J.H. 1998. 'Securing Compliance with the TRIPs Agreement after US vs. India. *Journal of International Economic Law* 1 (4): 585–601.

Reinert, E.S. 2009. *Emulation vs. Comparative Advantage: Competing and Complementary Principles in the History of Economic Policy*. Working Papers in Technology Governance and Economic Dynamics No. 25, Tallinn University of Technology.

Reis-Castro, L., and K. Henrickx. 2013. Winged Promises: Exploring the Discourse on Transgenic Mosquitoes in Brazil. *Technology in Society* 35: 118–128.

Reis, R., V. Terto, and C. Pimenta. 2009. *Intellectual Property Rights and Access to ARV Medicines: Civil Society Resistance in the Global South*. Rio de Janeiro: ABIA.

Ribeiro, L.L., and A.G. Carvalho. 2008. Private Equity and Venture Capital in an Emerging Economy: Evidence from Brazil. *Venture Capital* 10 (2): 111–126.

Roa-Rodriguez, C., and T. Dooren. 2008. Shifting Common Spaces of Plant Genetic Resources in the International Regulation of Property. *The Journal of World Intellectual Property* 11 (3): 176–202.

Rocha, G.M. 2002. Neo-Dependency in Brazil. *New Left Review* 16 (July–August): 5–33.

———. 2007. Celso Furtado and the Resumption of Construction in Brazil. *Latin American Perspectives* 156 (34): 1–28.

Rodriguez, A. 2008. *Knowledge and Innovation for Competitiveness in Brazil*. Washington, DC: World Bank Publications.

Rodrik, D., and A. Subramanian. 2004. *From 'Hindu Growth' to Productivity Surge: The Mystery of the Indian Growth Transition*. NBER Working Paper 10376.

Rodrik, D. 2013. The Perils of Premature Deindustrialization. *Project Syndicate*, October 11.

———. 2015. *Premature Deindustrialization*. Working Paper No. 20935. Cambridge, MA: National Bureau of Economic Research.

Rosendal, K. 1994. Implications of the US 'No' in Rio. In *Biodiplomacy: Genetic Resources and International Relations*, ed. V. Sánchez and J. Calestrous, 87–103. Nairobi: African Centre for Technology Studies.

Rudolph, L., and S. Rudolph. 1987. *In Pursuit of Lakshmi*. Chicago: University of Chicago Press.

Ryan, M.P. 1998. *Knowledge Diplomacy: Global Competition and the Politics of Intellectual Property*. Washington, DC: Brooking Institution Press.

Safrin, S. 2004. Hyperownership in a Time of Biotechnological Promise: The International Conflict to Control the Building Blocks of Life. *The American Journal of International Law* 98: 641–685.

Sahai, S. 2004. India Needs a Biotechnology Policy. *Frontline* 21 (10). Available at http://www.frontline.in/navigation/?type=static&page=flonnet&rdurl=fl2110/stories/20040521001708200.htm. Accessed 5 July 2016.

Sammut, S., and D. Levine. 2014. Accelerating Growth: Forging India's Bioeconomy. https://www.bio.org/sites/default/files/files/Burrill_AcceleratingGrowth_India-6-9-final.pdf. Accessed 19 June 2017.

Santilli, J. 2012. *Agrobiodiversity and the Law: Regulating Genetic Resources, Food Security and Cultural Diversity*. New York: Earthscan.

Sauer, S., and S.P. Leite. 2012. Agrarian Structure, Foreign Investment in Land, and Land Prices in Brazil. *The Journal of Peasant Studies* 39 (3/4): 873–898.

Scherer, F.M. 1993. Pricing, Profits and Technological Progress in the Pharmaceutical Industry. *The Journal of Economic Perspectives* 7 (3): 97–115.

Schuler, P. 2004. Biopiracy and Commercialization of Ethnobotanical Knowledge. In *Poor People's Knowledge*, ed. J.M. Finger and P. Schuler, 156–182. Washington, DC: World Bank.

Schwartz, R. 1988. Brazilian Culture. *New Left Review* 167: 77–90.

Schwartzman, S. 1995. *Science and Technology Policy in Brazil: A New Policy for a Global World*. Rio de Janeiro: Fundação Getúlio Vargas.

Scoones, I. 2002. Biotech Science, Biotech Business: Current Challenges and Future Prospects. *Economic and Political Weekly* 37 (27): 2725–2733.

———. 2005. *Contentious Politics, Contentious Knowledges: Mobilising against GM Crops in India, South Africa and Brazil*. Institute of Development Studies, Working Paper 256, Brighton UK.

Senker, J. 1996. National Systems of Innovation, Organizational Learning and Industrial Biotechnology. *Technovation* 16 (5): 219–229.

Sell, S. 1999. Multinational Corporations as Agents of Change: The Globalization of Intellectual Property Rights. In *Private Authority and International Affairs*, ed. A.C. Cutler, V. Haufler, and T. Porter, 169–198. Albany: State University of New York Press.

———. 2003. *Private Power, Public Law: The Globalization of Intellectual Property Rights*. Cambridge: Cambridge University Press.

Selwyn, B. 2011. Trotsky, Gerschenkron and the Political Economy of Late Capitalist Development. *Economy and Society* 40 (3): 421–450.

Shadlen, K. 2011. The Political Contradictions of Incremental Innovation: Lessons from Pharmaceutical Patent Examination in Brazil. *Politics & Society* 39 (2): 143–174.

———. 2016, forthcoming. *Global Change, National Responses: Social Coalitions and the Political Economy of Pharmaceutical Patents in Latin America*. Oxford: Oxford University Press.

———. 2017. *Global Change, National Responses: The New Politics of Pharmaceutical Patents in Latin America*. Oxford: Oxford University Press.

———., and E.M. Fonseca. 2013. Health Policy as Industrial Policy: Brazil in Comparative Perspective. *Politics and Society* 41 (4): 561–587.

Shaffer, G., M.R. Sanchez, and B. Rosenberg. 2008. The Trial at Winning at the WTO: What Lies Behind Brazil's Success. *Cornell International Law Journal* 41: 383–501.

Shaver, L. 2011. Access to Knowledge: From Development to Human Rights. In *Access to Knowledge in India: New Research on Intellectual Property, Innovation and Development*, ed. L. Shaver and R. Subramanian, 1–13. London: Bloomsbury.

Shubber, K., and N. Bullock. 2015. US Biotech Bull Run Sets Scorching Pace. *Financial Times*, July 17. Available at http://www.ft.com/cms/s/0/f33ffd50-2adb-11e5-8613-e7aedbb7bdb7.html#axzz3jpSMpX00

Silva, R.E., A.A. Amato, and M.R.C.G. Novaes. 2015. Funding of Clinical Trials in Brazil for the Development of New Drugs: Who Are the Sponsors? *International Journal of Clinical Trials* 2 (4): 75–84.

Silverstein, D. 1991. Patents, Science, and Innovation: Historical Linkages and Implications for Global Competitiveness. *Rutgers Computer and Technology Journal* 17 (2): 261–319.

Smith, C. 2000. China Rushes to Adopt Genetically Modified Crops. *New York Times*, October 7. Available at www.ag.ohio-state.edu/ipm/trans/10_071.htm. Accessed 25 January 2017.

Smith, N. 2007. Nature as Accumulation Strategy. *Socialist Register* 43: 1–21.

Smith, T. 2004. Technology and History in Capitalism: Marxian and Neo-Schumpeterian Perspectives. In *The Constitution of Capital: Essays on Volume One of Marx's Capital*, ed. R. Bellofiore and N. Taylor, 217–242. New York: Palgrave Macmillan.

———. 2010. Technological Dynamism and the Normative Justification of Global Capitalism. In *Political Economy and Global Capitalism: The 21st Century, Present and Future*, ed. R. Albritton, B. Jessop, and R. Westra, 25–41. London: Anthem Press.

Souza, V.S., and R.V. Santos. 2014. The Emergence of Human Population Genetics and Narratives About the Formation of the Brazilian Nation. *Studies in History and Philosophy of Biological and Biomedical Sciences* 47: 97–107.

Sridevan, P., P.M. Singh, N. Sabharwal, et al. 2014. *Draft National IPR Policy*. Department of Industry Policy and Promotion. Available at http://dipp.nic.in/English/Schemes/Intellectual_Property_Rights/IPR_Policy_24December2014.pdf. Accessed 16 July 2016.

Stiglitz, J. 1999. Knowledge as a Global Public Good. In *Global Public Goods: International Cooperation in the 21st Century*, ed. I. Kaul, I. Grunberg, and M. Stern, 308–325. New York: Oxford University Press.

Suchitra, M. 2012. India Has Approved Ratification of Nagoya Protocol. *Down To Earth*, October 9. Available at http://www.downtoearth.org.in/news/india-has-approved-ratification-of-nagoya-protocol--39280. Accessed 25 January 2017.

Sweet, C. 2013. The Political Economy of Pharmaceutical Production in Brazil. In *The New Political Economy of Pharmaceuticals Production, Innovation and TRIPS*, ed. H. Löfgren. Basingstoke: Palgrave.

Tregenna, F. 2016. Deindustrialisation and Premature Deindustrialisation. In *Handbook of Alternative Theories of Economic Development*, ed. E. Reinert, J. Gosh, and R. Kattel, 710–728. Cheltenham: Edward Elgar.

Tulchin, J.S., and H.A. Golding, eds. *Environment and Security in the Amazon Basin*. Washington, DC: Woodrow Wilson International Center for Scholars.

Ugolini, M.A. 1999. Grey Market Goods Under the Agreement on Trade-Related Aspects of Intellectual Property Rights. *Transnational Law* 12: 451–468.

Velez E. 2010. *Brazil's Practical Experience with Access and Benefit Sharing and the Protection of Traditional Knowledge*. Policy Brief Number 8, ICTSD, Geneva, Switzerland.

Viana, A.L., H.P. Silva, and I. Yi. 2015. *Universalizing Health Care in Brazil: Opportunities and Challenges*. UNRISD Working Paper 2015–8. Geneva: UNRISD.

Vielle, P. 1988. The State of the Periphery and Its Heritage. *Economy and Society* 17 (1): 52–89.

Visvanathan, S., and C. Parmar. 2002. A Biotechnology Story: Notes from India. *Economic and Political Weekly* 37 (27): 2714–2724.

Waldmeier, P. 2015. China Looks to Traditional Medicine as Tonic to Boost Growth. *Financial Times*, November 12. Available at https://www.ft.com/content/5b1969f0-7bce-11e5-a1fe-567b37f80b64. Accessed 2 April 2017.

Warman, A. 2003. *Corn and Capitalism: How a Botanical Bastard Grew to Global Dominance*. Trans. by N.L. Westrate. Chapel Hill: University of North Carolina Press.

Watal, J. 2001. *Intellectual Property Rights in the WTO and Developing Countries*. The Hague: Kluwer Law International.

Weiss, L. 2014. *America Inc*. Ithaca, NY: Cornell University Press.

Welch, J.R. 2015. Brazil's New Biodiversity Law. *Ethnobiology Letters* 6 (1): 216–217.

WHO. 2005. Intellectual Property Protection: Impact on Public Health 2010. http://www.who.int/medicines/areas/policy/AccesstoMedicinesIPP.pdf. Accessed 14 May 2016.

Wiebe, B., T. Hughes, and T. Pinch. 1987. *The Social Construction of Technological Systems*. Cambridge, MA: MIT Press.

Wilkes, G. 1984. Germplasm Conservation Toward the Year 2000. In *Plant Genetic Resources: A Conservation Imperative*, ed. C. Yeatman, D. Kafton, and G. Wilkes, 131–163. Boulder, CO: Westview Press.

Wilkinson, J., and S. Herrera. 2010. Biofuels in Brazil: Debates and Impacts. *The Journal of Peasant Studies* 37 (4): 749–768.

Wilson, P., and A. Rao. 2012. India's Role in Global Health R&D' Centre for Global Health R&D Policy Assessment (Washington, DC, USA). Available at http://healthresearchpolicy.org/sites/healthresearchpolicy.org/files/assessments/files/R4D%20-%20Indias%20Role%20in%20Global%20Health%20RD%20Final.pdf. Accessed 10 July 2016.

WIPO. 1988. Existence, Scope and Form of Generally Internationally Accepted and Applied Standards/Norms for the Protection of Intellectual Property, issued as GATT document MTN.GNG/NG11/W/24/Rev1. Geneva WIPO.

Wong, A.Y., and A.W. Chan. 2016. Genetically Modified Foods in China and the United States: A Primer of Regulation and Intellectual Property Protection. *Food Science and Human Wellness* 5 (3): 124–140.

Woo-Cumings, M. 1999. Chalmers Johnson and the Politics of Nationalism and Development. In *The Developmental State*, ed. M. Woo-Cumings, 1–31. London: Cornell University Press.

Yu, P.K. 2009. A Tale of Two Development Agendas. *Ohio Northern University Law Review* 35: 465–573.

Yu, P. 2013. The First Decade of TRIPS in China. In *China and Global Trade Governance: China's First Decade in the World Trade Organization*, ed. K. Zeng and W. Liang, 126–143. Abingdon: Routledge.

Zeller, C. 2007. From the Gene to the Globe: Extracting Rents Based on Intellectual Property Monopolies. *Review of International Political Economy* 15 (1): 86–115.

Index

V. Muzaka, *Food, Health and the Knowledge Economy*, Building a Sustainable Political Economy: SPERI Research & Policy, DOI 10.1057/978-1-137-59306-1